Palgrave Historical Studies in the Criminal Corpse and its Afterlife

Series editors
Owen Davies
School of Humanities
University of Hertfordshire
Hatfield, UK

Elizabeth T. Hurren
School of Historical Studies
University of Leicester
Leicester, UK

Sarah Tarlow
History and Archaeology
University of Leicester
Leicester, UK

Aim of the Series

This limited, finite series is based on the substantive outputs from a major, multi-disciplinary research project funded by the Wellcome Trust, investigating the meanings, treatment, and uses of the criminal corpse in Britain. It is a vehicle for methodological and substantive advances in approaches to the wider history of the body. Focussing on the period between the late seventeenth and the mid-nineteenth centuries as a crucial period in the formation and transformation of beliefs about the body, the series explores how the criminal body had a prominent presence in popular culture as well as science, civic life and medico-legal activity. It is historically significant as the site of overlapping and sometimes contradictory understandings between scientific anatomy, criminal justice, popular medicine, and social geography.

More information about this series at
http://www.springer.com/series/14694

Owen Davies · Francesca Matteoni

Executing Magic in the Modern Era

Criminal Bodies and the Gallows in Popular Medicine

Owen Davies
University of Hertfordshire
Hatfield, UK

Francesca Matteoni
Pistoia, Italy

Palgrave Historical Studies in the Criminal Corpse and its Afterlife
ISBN 978-3-319-59518-4   ISBN 978-3-319-59519-1   (eBook)
DOI 10.1007/978-3-319-59519-1

Library of Congress Control Number: 2017943494

Cover pattern: © Melisa Hasan

Printed on acid-free paper

This Palgrave Macmillan imprint is published by Springer Nature
The registered company is Springer International Publishing AG
The registered company address is: Gewerbestrasse 11, 6330 Cham, Switzerland

*Bah! What matters it? A man hanged, and then a rope
to be taken away, a scaffold to be unnailed, a corpse to be buried;
what does it all amount to?*

*Victor Hugo, Victor Hugo: A Life Related by One Who Has Witnessed It,
Vol. 2 (London, 1863), p. 214.*

CONTENTS

Contents

Introduction

Abstract This chapter sets out the aims and scope of the book, and provides contextual discussion on the history of public execution and post-mortem punishment in Europe from the eighteenth century onwards. It also explores the changing role of the European executioner over this period, and the sorts of people who became executioners.

Keywords Public execution · Post-mortem punishment · Executioners

The use of corpses in medicine and magic has been recorded dating back into antiquity. The lacerated bodies of Roman gladiators were used as a source of curative blood, for instance. In early modern Europe, a great trade opened up in ancient Egyptian mummies, plundered as a medicinal cure-all, and into the nineteenth century skull moss and oil of man (a distillation of human bones) were requested from chemists and druggists. Some of this unusual history has already been well told, notably by Richard Sugg in his book *Mummies, Cannibals and Vampires* (2011). This current study takes the subject into the modern era and focuses on the only corpses that continued to be popularly available for medical and magical usage—those of executed criminals.[1] However, we cannot attempt to understand the potency of the criminal corpse without considering its relationship to the executioner, the gallows, and the tools of execution, for all these accrued reputations for their healing and magical

© The Author(s) 2017
O. Davies and F. Matteoni, *Executing Magic in the Modern Era*,
Palgrave Historical Studies in the Criminal Corpse and its Afterlife,
DOI 10.1007/978-3-319-59519-1_1

properties. What was it about the act of execution that generated such beliefs, traditions and practices?

This history of execution medicine and magic begins in the age of the so-called Enlightenment, when centuries-old penal policies were being fundamentally questioned by the likes of the Italian criminologist Cesare Bonesana-Beccaria. His influential book *On Crimes and Punishments* (1764) attacked the use of torture and the death penalty on both ethical and practical grounds. Execution was principally a retributive act, he stated, and was patently not effective as a deterrent. The study ends in the early twentieth century, by which time most states across Europe, and the United States, with its adaptation of British common law, had either ended the practice of public execution or abolished capital punishment full stop.

Medical cannibalism and corpse magic are generally considered as premodern topics, ones that can only be understood in the context of the era of the witch trials, judicial torture, the adherence to ancient medical theories and pervasive illiteracy, or as aberrations that survived as 'superstitions' into the 'Age of Enlightenment' or 'long eighteenth century' as matters for antiquarians and folklorists. However, the aim of this book is to show how popular beliefs and practices regarding the executed and executions were not relics of an early modern past. They were adaptable to and were shaped by changing attitudes towards capital punishment. Indeed, as we shall see, in some respects, popular resort to the criminal corpse for magic and healing was inadvertently promoted and institutionalised by pragmatic and enlightened state penal policies.

Despite the significance of executions and executioners in the cultural and psychological relationship between populace and authority in modern-era Europe and America, it is difficult to piece together their histories in relation to the unorthodox, post-mortem histories of criminal corpses. Most insights are derived not from official records, but from sporadic, anecdotal, and often retrospective sources. Detecting the nuances of regional and local traditions and popular practices across Europe can only be an impressionistic exercise. Press reportage has been crucial to us, and this book could not have been written as it is without the search opportunities provided by the now vast collections of digitised newspapers. The valuable material collected by nineteenth- and early twentieth-century folklorists comes with its own interpretational problems. The fundamental challenge has been to use these sources to interpret what people thought or believed from the actions that were reported: distilling inner lives from physical expressions. Exploring the medical and magical use of the post-mortem criminal corpse and the execution environment becomes as much an exercise in intimate psychogeography as macrocosmic social and cultural history.

Public Execution and Post-mortem Punishment

During the eighteenth and nineteenth centuries, the morbid variety of non-military execution practices was significantly reduced. Death by drowning, boiling and burning had ended across much of Europe by the mid-1700s. The last cases of drowning in Amsterdam, for instance, occurred in 1730 for the crime of sodomy.[2] In Scandinavia and some German states, decapitation by axe or sword was still in use well into the nineteenth century. Indeed, it was adopted as a more humane method to replace hanging. The Polish penal code of 1818 meted out beheading by the sword as the standard mode of capital punishment, with only exceptionally heinous male criminals being hanged. In revolutionary France, the guillotine replaced the wide variety of punishments meted out under the *Ancien Régime*. Greece also adopted the guillotine in the 1830s, as well as practising execution by firing squad.

Sometimes certain types of crime were punished in specific ways, so in the German states of Prussia and Hesse- Kassel breaking on the wheel was still being employed into the 1830s to punish exceptionally heinous criminals, such as robbers who murdered their victims. An American traveller in Prussia in the 1820s described the two methods of breaking the body in shocked terms: 'The first is called the *Upper*, by which the head is broken first, and afterwards the breast and limbs ... The other is called the *Under*. The mode here is to break the limbs first, and afterwards the breast and head. The torture is thus prolonged.' In fact, it was usual for the criminal to be discreetly strangled with a cord just before the breaking began.[3] A few other types of aggravated death penalty continued to exist during the first three decades of the century. In 1835, for instance, the Assize judges of Mainz ordered that the murderer Margueret Jaeger should have her hand cut off just prior to her execution for the specific crime of parricide.[4]

Hanging had long been the main form of execution in Britain and Spain, though in the latter country, in 1822, the garrotte became the sole mode of operation. This was a form of strangulation that bypassed all the shortcomings of hanging techniques. It consisted of an iron collar, fixed to a post, with a heavy-handled screw or lever that could be tightened. Although it looked like an instrument of torture, it was thought, with reason, to be a more sure and humane method than hanging.[5] Other shifts in state practice occurred elsewhere. Hanging became the preferred means of execution in the Austro-Hungarian Empire from the mid-nineteenth century onwards, while in contrast hanging in Sweden, which was usually the punishment for commoners until the end of the eighteenth century, fell into abeyance during the next century when the vast majority of executions were by beheading with an axe or sword.

Post-mortem punishments such as dissection and gibbeting continued into the nineteenth century in some states. While the Netherlands abolished the practice in 1795, the last gibbeting in England took place in 1832. In the same year, the Anatomy Act ended the public dissection of executed murderers in Britain. The exposing of criminal corpses on the wheel ceased in Prussia in 1811, with most German states following suit if they had not already done so. In Norway, it was decreed, in 1832, that body parts and heads were to be removed from poles after 3 days instead of the previous practice of leaving them indefinitely in public view. The following year, the law was revised so that only the head was to be left on poles, and the practice was abolished entirely in 1842.[6]

Public civil execution ceased in most German states during the 1850s and 1860s, and in Britain and Austria in 1868. The last public executions (by beheading) in Sweden were in 1876, the last in Spain (by garrotting) in 1897. A number of states abolished the death penalty altogether for civilian crimes. Several German principalities and city states, such as Bremen and Hamburg, did this in 1849. Romania ended the death penalty in 1864, Portugal in 1867, and the Netherlands in 1870.[7] In Italy, the Zanardelli penal code of 1889 abolished the death penalty for civilian crimes. 'France, by contrast, initiated a very long game of hide-and-seek,' explains Paul Friedland, 'with officials desperately seeking to limit the visibility of executions on one side, and spectators equally desperate to see them on the other. Guillotines were exiled to the outskirts of town; raised platforms were outlawed in the hopes of limiting spectator visibility; executions were performed with little notice and at the crack of dawn.'[8] The last public execution in France was at Versailles in 1939, when German criminal Eugène Weidmann was guillotined before several hundred spectators.

So, as this brief overview suggests, against the background of the general diminution in the number of executions and of punishment as public spectacle, there remained considerable diversity of policy and public experience across Europe as states responded differently to the debates about capital punishment and the display of criminal corpses. These shifts and changes had obvious ramifications for the ways in which people were able to resort to the scaffold for cure, luck and protection. As we shall see, though, it was by no means a story of inexorable and inevitable decline of traditional practices from the early modern to the modern era of capital punishment. Even the ending of public executions did not stop popular desire for a piece of the action.

THE EXECUTIONER

As already noted, this study is as much about executioners as criminal corpses. The former drew power and influence from their relationship with the criminals they despatched and the subsequent curation of their corpses. As the 'embodiment of sovereign power', the executioner enacted the separation of the soul from the body as part of a secular and religious ritual, so it is not surprising that they accrued a popular thaumaturgic reputation. However, during the period concerned, the executioner's control of the scaffold was undermined by the state, and his grip on the curation of the criminal corpse loosened and was ultimately lost.

Despite their loss of influence, and then increasing invisibility due to the abolition of public execution, where executioners still operated they were high-profile figures in the public conscientiousness. Writing about the status of the French executioner in 1872, Maxime Du Camp observed: 'People no longer ask him for the grease from corpses to make love potions and mysterious unguents; but he remains nonetheless an obscure and much dreaded personage.'[9] Their names continued to evoke macabre curiosity and a degree of fear; they were figures people wanted to know about but not necessarily meet, bogeymen to frighten children. As a Hungarian writer reminisced in the mid-twentieth century, 'In my childhood the hangman's name was Michael Bali; and if a dangerous criminal was at large, people used to say, "Bali's rope will get him yet."'[10] In the age of the international press, they were of global celebrity interest and their deaths were widely reported. When the Austrian executioner Heinrich Willenbacher died in 1886, the British *Illustrated Police News* reported, for instance, 'In Willenbacher, the late hangman of Vienna, the natives of that effervescent city must have lost what is called a "type." He shuffled off this mortal coil in consequence of a sore throat, contracted while in the execution of a subsidiary profession. That is to say, he was dog catching at the time, for such was his unofficial calling.[11] There was sometimes an ironic element in such reports, along the lines of the biter has been bitten, so the death of the Moravian public executioner, a man named Bott, was reported in 1884 because it was suspected he had been murdered in a revenge attack.[12]

In some states, the job of executioner remained a hereditary affair. The most well-known example is the Sanson family in France, who held the position of Paris executioner from 1688 to 1847. Karel Huss (1761–1838), the last Bohemian executioner, followed in the footsteps of his father and uncle. Alois Seyfried, the Bosnian hangman who became the first executioner of Yugoslavia, was related to Willenbacher.[13] The once widespread tradition of appointing executioners from convicted criminals facing death penalties was outlawed in Sweden in 1699 but continued elsewhere.[14] In England, the York executioner, William Curry, who held the office between 1802 and 1835, was a recidivist sheep-stealer who had been sentenced to death twice during his criminal career before opting to become a hangman.[15] The notorious Greek executioner and former convict Bekiaris, who died in 1903, murdered his mistress while in post, leading to the titillating international news that an executioner faced being executed, but clearly could not do the job himself. In southeastern Europe, the tradition continued of employing another 'outcast' group, the gypsies, men such as Bulgarian hangman Hussein Jasara.[16]

The profession had its fair share of erudite gentlemen. Karel Huss was an archetypal Enlightenment figure. He was a historian, ethnographer, philosopher and collector of curiosities, including executioners' swords. He was already performing executions in his late teens, and when his uncle, the executioner of the popular spa town of Cheb, retired, Karel took over the formal position. He put the social exclusion the post brought with it to good use by concentrating on his educational, cultural and intellectual improvement and moving in enlightened circles, which, in turn, brought numerous cultured visitors to his home, such as the famed German poet Goethe. One of Huss's works was a critique of local popular 'superstition' based on personal experience, entitled *O pověrách*, or *Vom Aberglauben* in German, which included folklore regarding executioners.[17]

In a few places, the title of executioner was an official position that did not actually require the holder to carry out executions. This peculiar state of affairs pertained to Caspar Frederik Dirks, a physician and apothecary who, in 1780, applied for and obtained the role of *skarpret-ter* (executioner) on the remote Danish Baltic island of Bornholm. Dirks, who was aged 55 at the time, had found it difficult to make ends meet from his medical business on the island, and traded in French brandy to supplement his income. The annual executioner's salary of 100–120 *riks-daler* was attractive, particularly considering that executions were rare on the island. His application also included the offer to supervise infectious diseases amongst the population. He got the job and as part of the deal he paid an annual retainer to a neighbouring *skarpretter* to conduct any hangings. For 11 years, all went well in the absence of any executions. However, in 1791, one was required, and the man he had paid on retainer refused to do it, so Dirks had to pay 200 *riksdaler* to hire the Copenhagen *skarpretter* to fulfil royal justice. Dirks continued in post until his death in 1802, when his widow applied to take over the role. This would have made her the only female executioner in Europe, but not surprisingly her application was denied.[18]

No matter the background of executioners, and the diminution of their influence: the stigma of the 'dishonourable trade' remained. In revolutionary France, the citizen rights of executioners were debated in 1789 as part of the Declaration of the Rights of Man. The citizen–aristocrat Clermont-Tonnerre spoke up for their inclusion. 'The executioner simply obeys the law,' he said. 'It is absurd that the law should say to a man: do that, and if you do it, you will be abhorrent to your fellow men.' The *abbé* Maury took a more familiar view: 'The exclusion of public executioners is not founded on a mere prejudice. It is in the heart of all good men to shudder at the sight of one who assassinates his fellow man in cold blood. The law requires this deed, it is said, but does the law command anyone to become a hangman?'[19] Over a century later, Michael Bali, one of the last European hangmen to sell pieces of his ropes, wrote an indignant letter to the president of the Hungarian National Assembly after reports that the phrase 'the hangman is your friend' had been bandied about as a political insult during debates. Bali stated his trade was 'as honourable and useful as that of judges, lawyers, ministers or kings, why should the old superstition hold in modern times that the hangman's profession is disgraceful, abominable and loathsome? My friends are all perfect gentlemen and any member of parliament can consider it only flattery when he is called my friend.'[20] Still, despite the opprobrium that came with the job, when the position of executioner became vacant or was openly advertised, there was usually no shortage of applicants. When the post of Swedish executioner was advertised in 1883, 28 people applied, including a butcher, baker, coppersmith, policeman and several soldiers. Thirteen people applied to fill the position of Madrid executioner following the death of Francisco Castellanos in 1894, including a barrister and an ex-sergeant major. When, in England, a vacancy for hangman was advertised in 1902 to replace the previous incumbent, William Billington, over 300 applications were received.[21]

***

While much of this book is concerned with the afterlife of criminal corpses, and their custodians, in order to understand the connections between criminality, identity, morality and the body, and their therapeutic and magical value in society and culture, we need to start by considering the potency and power of the living criminal body. What relationship, if any, was there between the healing and protective properties of the criminal corpse and its previous sentient state? Was the power of the body activated or enhanced by the act of execution, or did the living criminal already bear the signs of his or her corporeal value in death?

NOTES

1. The research for this book was kindly funded by the Wellcome Trust as part of the 'Harnessing the Power of the Criminal Corpse' project. We are very grateful to the Wellcome Trust for their support. We would like to thank our fellow colleagues on the project, Sarah Tarlow, Elizabeth Hurren, Pete King, Richard Ward, Zoë Dyndor, Shane McCorristine, Floris Tomasini, Rachel Bennett and Emma Battell Lowman. It has been a real pleasure to share and exchange ideas with them over the years of the project. And finally thanks to Sarah for inviting us to be part of the team in the first place, and for leading the project in such a genial and effective way.
2. Pieter Spierenburg, *The Spectacle of Suffering: Executions and the Evolution of Repression* (Cambridge, 1984), p. 71.
3. Henry Edwin Dwight, *Travels in the North of Germany: In the Years 1825 and 1826* (New York, 1829), p. 214.
4. *The Essex Standard*, 10 April 1835.
5. Ruth Pike, 'Capital Punishment in Eighteenth-Century Spain', *Histoire sociale-Social History* 18 (1985) 383.

6. David Garland, *Peculiar Institution: America's Death Penalty in an Age of Abolition* (2010), pp. 102–103; Richard J. Evans, *Rituals of Retribution: Capital Punishment in Germany, 1600–1987* (Oxford, 1996), pp. 226–227; John Pratt and Anna Eriksson, *Contrasts in Punishment: An Explanation of Anglophone Excess and Nordic Exceptionalism* (Abingdon, 2013), p. 98.

7. Jos Monballyu, *Six Centuries of Criminal Law: History of Criminal Law in the Southern Netherlands and Belgium (1400–2000)* (Leiden, 2014), pp. 157–158.

8. Paul Friedland, 'The Last Public Execution in France', Oxford University Press Blog, 17 June 2012, http://blog.oup.com/2012/06/last-public-execution-france-17-june-1939/. More generally, see Paul Friedland, *Seeing Justice Done: The Age of Spectacular Punishment in France* (Oxford, 2012).

9. Cited in Richard D.E. Burton, *Blood in the City: Violence and revelation in Paris, 1789–1945* (Ithaca, 2001), p. 102.

10. Bela Fabian, 'Lynching Soviet Style', *The American Legion Magazine* 69 (1960) 38.

11. *Illustrated Police News*, 3 April 1886.

12. *Western Times*, 2 January 1884.

13. Arthur Isak Applbaum, 'Professional Detachment: The Executioner of Paris', *Harvard Law Review* 109 (1995) 458–486; Frédéric Armand, *Les bourreaux en France: Du Moyen Age à l'abolition de la peine de mort* (Paris, 2012), pp. 161–172; 'Serbia against Capital Punishment', http://www.smrtnakazna.rs/en-gb/topics/theexecutioners/aloisseyfried.aspx; Hermann Braun, *Karl Huss: Scharfrichter und Folklorist* (Marktredwitz, 1977).

14. Finn Hornum, 'The Executioner: His Role and Status in Scandinavian Society', in Charles H. Ainsworth (ed.), *Selected Readings for Introductory Sociology* (New York, 1972), p. 72. Reprinted from the *Graduate Sociology Journal of the University of Pennsylvania* 4 (1965) 41–53.

15. James Bland, *The Common Hangman: English and Scottish Hangmen Before the Abolition of Public Executioners*, 2nd ed. (Westbury, 2001), pp. 107–110.

16. Widely reported in the press, though not always accurately. See, for example, *San Antonio Light*, 9 August 1925; *Pittsburgh Press*, 18 October 1932.

17. Hazel Rosenstrauch, *Karl Huß, der empfindsame Henker. Eine böhmische Miniatur* (Berlin, 2012); Braun, *Karl Huss: Scharfrichter und Folklorist*; John Alois (ed.), *Die Schrift 'Vom Aberglauben': nach dem in der fürstlich Metternischen Bibliotek zu Königswart befindlichen Manuscrkipte* (Prague,

1910), p. 21. There was a trade in fake executioners' swords during the nineteenth century; Vilém Knoll, 'Executioner's Swords—their Form and Development: Brief Summary', *Journal on European History of Law* 3 (2012) 158–162.

18. Kristian Carøe, *Bøddel og Kirurg* (Copenhagen, 1912), pp. 47–51; http://www.a-apoteket.dk/1171/historie.
19. Applbaum, 'Professional Detachment', 463–464.
20. *Sunday Oregonian*, 23 October 1921.
21. Hornum, 'The Executioner', p. 72; *York Herald*, 10 April 1894; *Cheltenham Gazette*, 1 March 1902.

Criminal Bodies

Abstract This chapter examines historic views on the potency, power and agency of the living criminal body in the early modern and modern periods as a way of understanding the potency of the criminal corpse. The main section of the chapter focuses on the witch as the most powerful of living criminal bodies. There is discussion on phrenological interpretations of criminality and the work of Cesare Lombroso on the 'born criminal'. The meaning of cruentation, or the ordeal by bleeding corpse, is also explored.

Keywords Witch · Phrenology · Humours · Bleeding corpse Lombroso

In the medieval and early modern period, it was widely thought that God left his imprints on all living things, and it was an aspect of natural magic for humans to try and interpret their meaning to understand better the world He had created. With regard to human bodies, this meant that the lines on the hand, the wrinkles on the forehead, the shape of the nose, the colour of hair, the number of moles and other visible bodily features, signified how God moulded each person and imbued him or her with an individual character, identity and destiny. This art or science of physiognomy drew on concepts from the ancient world that expounded all-encompassing theories regarding the interconnectedness

© The Author(s) 2017
O. Davies and F. Matteoni, *Executing Magic in the Modern Era*,
Palgrave Historical Studies in the Criminal Corpse and its Afterlife,
DOI 10.1007/978-3-319-59519-1_2

of matter. The doctrine of signatures, which governed much of herbal medicine, observed, for instance, that plants, animals and objects that resembled parts of the body were imbued with healing properties appropriate to that body part. Thus, liverwort was used for liver complaints. In Christian terms, God left such clues throughout the natural world to aid humankind. Physiognomy was also tied up with Galenic humoral theory, in other words, the notion that the body was governed by four humours—black bile, yellow bile, blood and phlegm. Imbalances between these humours caused illness and behavioural problems, as well as hot and cold temperaments, which could only be cured by restoring a healthy equilibrium. The notion that criminals had 'bad blood' derived, for example, from the idea that they were prone to excess bile. In other words, they had bilious constitutions. The traits of potential criminality—or, at least, the humoral passions and characteristics symptomatic of criminal behaviour—might be externally observable. Therefore, people with red hair, which in Galenic terms denoted a hot, choleric temperament, were considered more prone to violence, and a monobrow denoted a dangerous person. Such associations remain in the popular consciousness even today.[1]

Renaissance anatomists believed that through their dissections of executed criminals they found physical evidence that the criminal body could also be *anatomically* different from normal bodies, just as saints' bodies exhibited signs of divine influence. As Katharine Park has noted, 'the deeds of both were assumed to be supernaturally inspired, whether by God or the Devil.' Thus, when Florentine physician Antonio Benivieni (1443–1502) dissected a notorious thief, he found what he believed to be hair covering his heart. This, he concluded, was due to a particularly hot complexion.[2] Advances in anatomical science over the next two centuries rendered such interpretations obsolete, but the ambition to be able to identify the signs of innate criminality in and on the human body did not go away. The search was now on for the biological causation of crime, rather than for signs of divine influence.

The late eighteenth- and early nineteenth-century anatomy schools had a regular supply of criminal corpses, and for some, influenced by early psychiatry, this opportunity enabled the study of the relationship between brain development and criminality. The most controversial exponent was the German physiologist Franz Josef Gall (1758–1828), who was the first to distinguish clearly between the grey and white matter of the brain. Gall's studies led him to believe that different parts or 'organs' of the brain controlled different character and behavioural traits; their lesser or greater size determined their expression through action and thought in everyday life. Gall believed, furthermore, that the skull developed in relation to the size of the different parts of the brain. This scientific idea that the bumps on the skull reflected the shape of the brain and, therefore, could be used to determine the character of the individual led to the pseudoscience of phrenology. Gall and the phrenologists identified two particular areas diagnostic of criminality: the organs of destructiveness and combativeness. Gall had originally identified an 'organ of murder', having found protuberances in the same place on two murderers' skulls, but his influential disciple Johann Spurzheim (1776–1832) disliked naming an organ 'according to its abuse' and so labelled it according to the propensity for destructive behaviour.[3] Gall and Spurzheim's analysis of the skulls of unwed mothers accused of infanticide apparently revealed, furthermore, that 25 out of 29 had a weak 'organ of love of children'. The implications of all this were deeply provocative from a religious point of view, as they challenged the fundamental link between sin and criminal behaviour. Gall was accused of undermining the unity of the soul, free will and Christianity itself. His publications were placed on the Catholic Church's *Index* of forbidden books.[4]

The use of phrenology as a predictive branch of criminology was also controversial. It stood to reason that if criminals could be identified phrenologically, then one could anticipate, control and prevent crime. The potential was explosive for penal policy. Writing in 1836, the Procurator Fiscal of Lanarkshire, George Salmond, thought that the new science could lead to 'the better classification of criminals confined before and after trial, to the selection and treatment of convicts, and even to the more certain identification of such criminals as might effect their escape from justice or confinement.'[5] But for every murderer's skull that seemed to confirm criminal tendencies, there were others that did not match. William Saville was hanged in Nottingham in 1844 for the murder of his wife and three children, yet a phrenological study of Saville concluded in frustration: 'there was nothing in the posterior part of the head which attracted particular attention. The organs of Destructiveness were not in the least protuberant. Combativeness and Amativeness were moderate. Now, what are we to say to all this? As an individual, I feel quite confounded.'[6] However, it was investigations inside the skull that thoroughly undermined the theory. From early on, fellow anatomists critiqued Gall's theories and those of his followers. The American surgeon T. Sewall stated categorically in 1837: 'the division of the brain into phrenological organs is entirely hypothetical; it is not sustained by dissection'. Because the early phrenologists had used comparative animal anatomy to draw up the typology of the various organs of the brain, referring to leonine, canine or dove-like qualities, for example, one initial advocate of Gall's work, John C. Warren, professor at the Massachusetts Medical College, decided to anatomise the head of a lion to confirm that the organs of combativeness and courage were unduly large. However, he found instead that they were no bigger than the corresponding organs of phrenologically inoffensive sheep.[7]

One sceptic wrote in the 1830s, 'phrenology has more lives than any cat, or it could not have survived till now, perplexing weak minds, though supported by very clever ones.'[8] Indeed, while phrenology eventually fizzled out of medical discourse during the mid-nineteenth century, it inspired the new discipline of criminal anthropology, and the notion of the 'born criminal' espoused by the influential Italian criminologist Cesare Lombroso (1835–1909). He and others measured, weighed and analysed hundreds of living and executed criminals to draw up a scientific typology of characteristics. Some features were behavioural consequences, such as a higher frequency of wrinkles—the criminal being more prone to cynical laughter—but other features were considered inherent, biological throwbacks to a more primitive stage of humanity and arrested development. Still, Lombroso had to admit that only 40% of convicted male criminals he analysed bore a characteristic criminal feature, and even less a combination that could convincingly *predict* criminal behaviour.[9] The criminal anthropologists also looked inside the body. It was thought in Italian folklore that the absence of blushing was a sign of a dissolute life, so blood-flows in different parts of the criminal body were measured and the dilation of blood vessels in the face analysed, leading Lombroso to confirm that criminals were not prone to facial flushes. His studies 'showed' that of a sample of 122 female criminals, 79% of murderers, 82% of infanticides and 90% of thieves could not blush.[10] Lombroso and his ilk were, in some respects, trying to confirm centuries-old received wisdom that had been underpinned by humoral theory. He also observed, for instance, that red-haired people were disproportionately more prone to criminality, particularly crimes of lust.

The eighteenth- and nineteenth-century attempts to identify and classify born criminals or, at least, indicate individual's propensity for crime, were framed as scientific endeavours, and some good science as well as a lot of very bad science emerged from the work of phrenologists and criminal anthropologists. But in one sense, they were driven by the same venerable desire as the natural magicians centuries before, that is they were trying to understand how nature not nurture stigmatised the criminal body.

THE WITCH

While the anatomists, phrenologists and criminologists carried out their probing, prodding and cutting of criminals and criminal corpses in the name of science, for very different reasons the common people were doing the same to the bodies of the most feared of living criminals—witches.

While the average criminal was generally a common person led to commit crime by a variety of circumstances and passions, the witch was an extraordinary evildoer, a supernatural criminal who could kill with a look or mere touch, who could transform into a cat or a hare better to perform his or her malicious acts of envy. Although by the late eighteenth century the crime of witchcraft was no longer recognised in European law (except for a few legal anomalies), fear of witchcraft was still widespread. Dozens of accused witches were murdered by mobs or individuals, and thousands were abused, ostracised and assaulted long after the witch trials ended.[11]

There was a rich folklore concerning what witches looked like and how they could be identified. From the early woodcut depictions in sixteenth-century witch-trial pamphlets to the folklore records of the nineteenth century, the body of the stereotypical witch was thought to bear physical traits of their criminality: some through birth, such as sharp prominent noses and squinting eyes; some characteristics engendered through age, such as a gobber tooth and hunched back.[12] As we have seen, according to Galenic medicine unbalanced humours caused sickness: an old woman, whose body ceased being cleansed through the menstrual process, became herself diseased and contagious. Something happened to ageing female bodies that relocated them to a marginal, aggressive zone. The horror of decay did not spread from the image of the wizened body, but from the grotesque and fetid internal state dependent on the functioning of blood. Weak and corrupted blood flowed towards disease and death and became the gateway to demonic

forces that disrupted the humoral balance. Toxicity was not merely physical: spiritual factors contributed to it and it expressly manifested itself
in melancholy, which resulted from the abundance of black bile, thick-
ened blood or corrupted, stagnant humours. These substances reached
the brain, sickening the mind and infecting the faculty of imagination,
which then became easy prey for the manipulative works of the Devil.
He took advantage of bad humours to interfere with the fantasies of the
witch, using her as a vehicle to attack the emotional environment of the
community. According to the ardent witch believer, Joseph Glanvill,
witches' familiars 'breath'd some vile vapour into the body of the Witch'
that tainted her blood and spirits with a 'noxious quality', which, in turn,
infected her imagination.[13] So body and imagination mutually influenced
each other through blood. In the words of Robert Burton, the author of
The Anatomy of Melancholy:

> For as the Body workes upon the Mind, by his bad humors, disturbing
> the Spirits, sending grosse fumes into the Braine; and so *per consequens*
> disturbing the Soule, and all the faculties of it, with feare, sorrow &c. ...
> so on the other side, the Minde most effectually workes upon the Body,
> producing by his passions and perturbations, miraculous alterations, as
> Melancholy, Despaire, cruell diseases, and sometimes death it selfe.[14]

This explanation provided a scientific basis for witchcraft *confessions*. But
even the likes of Reginald Scot, Elizabethan critic of 'witch-mongering',
also thought that it was possible for people to enchant or 'fascinate' others through the eyes through similar processes: 'For the poison and disease in the eie infecteth the aire next unto it, and the same proceedeth
further, carrieng with it the vapor and infection of the corrupted bloud:
with the contagion whereof, the eies of the beholders are most apt to be
infected.' Thereby postmenopausal women could wittingly and unwit-
tingly affect people through the 'vapours' which rose from their polluted
bodies and came out from the eyes and spread contagion into others:
'as these beames & vapors doo proceed from the hart of the one, so are
they turned into bloud about the hart of the other: which bloud disa-
greeing with the nature of the bewitched partie, infeebleth the rest of his
bodie'.[15]

The idea of the witch's body as a walking source of emanating pollution was grounded in the notion of the porosity and permeability of
human and animal bodies, and the consequent dangers of contamination

from fluids such as blood, urine and milk. In a number of eighteenth-century Swedish church court prosecutions we find, for example, accusations that witches turned cows' milk into blood. This was sometimes thought to be the result of witch-hares, which were made from witches' blood, sucking the cows' teats. The pure milk was contaminated as a consequence. This led to the notion of bewitched butter bleeding when cut. In the early 1770s, for instance, a widow named Anna Andersdotter suspected one Karin Mansdotter of bewitching her cows' milk. She took some butter made from the contaminated milk to a local cunning woman, who confirmed that the butter was 'full of blood and bloody pus'.[16]

Just as Lombroso had to admit that most criminals did not exhibit a complete set of criminal physical traits, so many accused witches, whether in the early modern or modern periods, did not fit the stereotypical image. Some were male, many were young women, and some were described as attractive. Therefore, we need to be careful about conflating contemporary demonological and medical theories regarding the witch's body with the complex social and emotional factors that actually led to people being accused of witchcraft. Just because you looked like a witch did not mean you were believed to be one.

Some attributes of the witch's body were accrued rather than being innate or consequent upon age and gender. In other words, a normal body was corrupted and marked as such by the Devil or his minions, every goodness inside the person was exchanged for the power to harm. Demonologists such as Jean Bodin and Jean Boguet attested that at sabbats the Devil placed his claw on the forehead of the witch to take away the power of the holy chrism and of the baptism. He might also ask witches to sign a pact with their own blood, as a symbol of loyalty, in the manner of ancient oaths. In his *Compendium Maleficarum* (1608), the Italian priest Francesco Guazzo stressed that of the witch's bodily goods the Devil claimed the blood.[17] Through the pact, the Devil impressed a seal on the human body, which invisibly transfigured it, turning it into a criminal body that needed to be abused and destroyed. Early modern confessions from Scotland, Denmark and the Basque region illustrate the demonological notion that the Devil physically assaulted the witches he recruited, biting, scratching, or painfully licking their bodies, provoking an injury that had permanent effects.[18] But how did this mark appear? And how did it influence the perception of the witch? According to the demonologists, the mark was usually well hidden, often close to the sexual organs, though men also bore it on the eyelids, the arm-pit or the shoulder, while women could have it on their breasts. More importantly, it was insensible and could not bleed even if pierced, and could be confused with a range of skin marks and imperfections.

The English witch received the mark in the form of a teat, an extra nipple, or a dark spot, from which a devilish familiar spirit drank blood. The familiar, which usually appeared in the form of a dog, cat, mouse or other small animal, had a symbiotic relationship with the witch. Just to mention a few famous examples, we can look at the accusations from the end of the sixteenth century. At Chelmsford in Essex, Elizabeth Francis confessed to keeping a familiar in the form of a spotted cat named Satan: 'Every time that he did anything for her, she said that he required a drop of blood, which she gave him by pricking herself, sometime in one place and then in another, and where she pricked herself there remained a red spot which was still to be seen'. The same familiar, changing its form from cat to toad and then to a black dog, was thought to be kept by Agnes and Joan Waterhouse, mother and daughter, who gave it blood by pricking their hands or faces and putting the substance directly into its mouth.[19] In 1579, at Windsor, other witches confessed to giving blood, sometimes mixed with milk, to their familiar spirits, taking it from 'the flank' and 'the right-hand wrist'.[20] At the famous trial at St. Osyth, Essex, in 1582, the 8-year-old Thomas Rabbet testified that his mother, Ursula Kemp, had four familiars, a grey cat called Titty, Tiffin, a white lamb, Piggin, a black toad, and Jack, a black cat, which, 'in the night-time will come to his mother and suck blood of her upon her arms and other places of her body'.[21] Descriptions of familiars changed through the seventeenth century, under the increasing influence of demonologi-cal beliefs about the Devil's mark. While they became more ethereal and impish in form and appearance, their sucking habits developed a sexual connotation, the teat being located 'a little above the fundiment', or 'in the secret parts' of the accused.[22]

Blood as a means of exchange, contagion and empowerment is the key element to understanding how the witch's criminal body affected the surrounding human landscape. Taken away by the Devil at the moment of the pact, as illustrated by Italian cases, it was compensated for from witches' young victims. According to the philosopher Pico Della Mirandola, the witch killed infants by drying up their blood. In a treatise written at the end of the sixteenth century, it was explained that mid-wives could sometime become witches, bewitching children to death, hitting and wounding their heads to suck the blood and the breath out.[23] During 1539, the Modenese witch Orsolina la Rossa di Gaiato confessed to wasting small children by sucking 'their blood from under the nails of their hands or of their feet, or else from their lips', and then preparing a *focaccia* with the congealed fluid.[24] In 1540, the accused witch Cecca confessed to having ridden on a he-goat with a friend to the house of Francesco Collavoli in San Miniato, a village close to Florence, where they sucked the blood from the left breast of Collavoli's little daughter, provoking her sickness and death within days.[25] These examples characterize the witch as a stealer of life force. Sickness caused by witchcraft was an act of theft and a struggle between two bodies: one healthier and younger, the other driven by an otherworldly aggressive thirst. When blood, the vital element, was exchanged between the Devil and the witch, the latter was granted a form of criminal power, which she actively exercised as a living vampire. However, the witch's blood itself was a nourishing, enriching substance for supernatural allies.

THE WITCH'S BLOOD AS A CURE

So far, we have been looking at early modern sources to try and understand how intellectuals conceptualised and explained the agency of the witch's criminal body according to religious and scientific thought at the time. However, the extent to which any of these ideas were understood in popular cultures is difficult to gauge. What is clear from the actions of people, though, is that witch's blood was thought to be inextricably linked to certain types of bewitchment. The act of witchcraft established a physical and spiritual relationship with the victim's body. To break the spell one could, therefore, attempt, in turn, to afflict the body of the witch. This sometimes manifested itself as a brutal physical assault, beating the witch into removing the spell. Harmful rituals of sympathetic magic, such as the use of witch bottles, were performed to cause witches excruciating pain at a distance. However, the ritualistic practice of drawing blood from a witch in early modern and modern England leads us back to notions of bodily imbalance and sympathy, and the witch's criminal body as liminal matter between the spirit and the physical world. There are numerous cases recorded from the era of the witch trials.[26] It was not a fail-safe therapy, though, because the blood did not always flow from the witch as it should. Thus when, in Norfolk in 1617, Edmund Newton attempted to draw blood from Mary Smith, the nails he used 'turned like feathers'.[27] When, in mid-seventeenth-century Yorkshire, one Richard Brown accused the suspected witch, Elizabeth Lambe, of drawing 'his heart's blood from him', he concluded that 'if he could draw blood of her, he hoped he should amend'. He then scratched her until the blood ran.[28] We can interpret Brown's thinking in terms of humoral balance, but in popular cultural terms he was probably guided by the near universal concept of limited good. This was the notion that there was a finite amount of everything, and that someone's loss was someone else's gain. If a cow dried up then a neighbour's cow must be producing more abundantly. Witches were supernatural agents in this, taking away the substance or replacing it with a polluted one and repurposing it for themselves.

While intellectual discourses on the polluting and liminal nature of the witch's body ceased by the early eighteenth century, the scratching of witches to draw blood continued into the nineteenth century and beyond. In 1846 in Appledore, Devon, Roger Fursdon and his daughter, who was thought to be bewitched, assaulted the 14-year-old Richard Evans with a knife and then a pin. For the same reason, in 1852 at Norwell, Nottinghamshire, Ann Williamson was scratched with a darning needle by Thomas Freeman. Two of his daughters had fallen ill and the first one, who was 'reduced to a complete skeleton', had frequently uttered the name of the old woman.[29] The level of assault was often carefully judged to draw blood but not to cause substantive wounding, as a Scottish prisoner from Tain explained in the 1840s:

> People believe in my neighbourhood that if anyone gets blood from a witch she can do them no more harm, and that is the reason I cut M. with my penknife, but I held the knife so that it might go into her as short a way as possible. All I wanted was to get blood. I was not the first person who wanted to draw blood from her. Those who advised me to cut her told me that if I did not she would drown me and the rest who were in the boat.[30]

In the numerous cases from the nineteenth and early twentieth century, there is no mention of the blood exchange indicated by Richard Brown above, nor hints of humoral or limited good notions. There is a general sense, of course, of health taken by the witch and health restored by the drawing of blood. It is possible that scratching had become received practice, while the rationale behind it was largely lost in popular belief.

What is clear is the apparent efficacy people attributed to the action. 'I have drawn your blood and broke the spell, and now you have no more power over him', and 'it was a lucky scratch for me', are representative statements from nineteenth-century cases. In 1867, John Davis from Stratford-upon-Avon was condemned to 18 months' hard labour for having assaulted Jane Ward, but he was happy because, he asserted, the witch had no more power over him, because he had got her blood.[31]

In north-eastern Europe, we find a distinct tradition of scratching or beating witches to draw blood in order to drink it as a cure. In a widely reported court case from Sztum (Poland) in 1880, it was heard how the Prussian cunning man, Dr. Kotlevski, had treated a woman suffering from epilepsy. He claimed she was possessed by four devils conjured into her body by a local witch. Three of the devils he boasted to have exorcised with his magic, but the fourth proved stubborn, he said, so he went to the suspected witch's cottage and attacked her with a heavy cudgel to draw blood and collect it for his patient. A neighbour intervened and Kotlevski was arrested. In Schonbeck, West Prussia, in 1883, the daughter of a cabinet-maker was recommended to ingest the blood obtained by pricking the finger of a supposed witch in order to be cured of the 3-year-long sickness she attributed to witchcraft. Seven years later, another prosecution ensued in Prussia when a well-to-do gentleman drew blood from the arm of an accused witch to give his bewitched wife to drink. Finally, in 1907, in the Polish village of Wieliszew, not far from Warsaw, Marya Zhroh, a farmer's daughter, attributed her illness to the witchery of her neighbour, Josephine Zlolkow. Following the victim's request, some villagers assaulted Zlolkow, collecting the blood flowing from her nose and ears and giving it to Marya to drink. She, then, like many witch-scratchers, affirmed to feel suddenly better.[32]

CORPSES, EVIDENCE AND FEELINGS

The witch was not the only living criminal body that engendered spiritual or magical responses. While in demonological thought the witch's criminality depended upon interaction with demonic entities, other criminal bodies generated supernatural or miraculous manifestations through the disruption of the blood and the soul, rather than the emanation of power or intrinsic potency. This is illustrated by the concept of cruentation or the ordeal by touch, whereby it was thought that the corpse of a murdered person would bleed when touched by the murderer.

In medieval theological terms, as Sara Butler observes, blood was the vessel containing the soul and so cruentation was 'easily understood as the soul speaking after the body has lost the capacity'.[33] Sometimes, the mere presence of the guilty would cause the blood to flow.

The practice of corpse-touching, and religious, legal and medical questions regarding it, seeped into the age of the witch trials and cropped up in the growing debates about the continuation of miracles beyond the biblical age.[34] In his *Daemonologie* (1597), King James I and VI noted that 'in a secret murther, if the deade carcase be at any time thereafter handled by the murtherer, it wil gush out of bloud, as if the blud wer crying to the heaven for revenge of the murtherer, God having appoynted that secret super-naturall signe.'[35] In considering the matter, the seventeenth-century physician and witchcraft sceptic, John Webster, argued that the moment of death did not always correspond to the departure of the soul:

> the Soul being yet in the Body, retaining its power of sensation, fancy and understanding, will easily have a presension of the murderer, and then no marvail that through the vehement desire of revenge, the irascible and concupiscible faculties do strongly move the blood, that before was beginning to be stagnant, to motion and ebullition, and may exert so much force upon the organs as for some small time to move the whole body, the hands, or the lips and nostrils.[36]

Webster explained the relationship between the body and the soul in terms of a divine love that violent death brutally interrupted: thus, the soul, still present in the corporeal dimension, claimed and obtained revenge through the sweating of the vital fluid.[37] Blood enlivened the residual vitality of the corpse and at the same time it constituted an instrument of the soul to re-establish a moral order, broken by criminal actions. Little was said, though, about the agency of the murderer in this relationship. But it clearly was essential.

Cruentation remained a judicial test into the eighteenth century in some parts of Europe, although it was increasingly held to be mere superstition by coroners and the judiciary. Eighteenth-century editions of Michael Dalton's influential guide for magistrates, *The Countrey Justice* (1630), continued to cite cruentation as a cause of suspicion in murder cases.[38] Into the nineteenth century, the ordeal remained a popular judicial reflex, though not an official one, in response to suspected murders. When, in 1827, a 15-year-old boy named James Urie was found drowned near Wandsworth in suspicious circumstances, the body was fished out and several people were asked to touch the face of the corpse. When a boy named Taylor did so, blood was seen to leak out of Urie's nose. This caused much suspicion in the neighbourhood, but no prosecution was sought for Urie's death. Three years later, in October 1830, William Edden, a gardener of Thame, Oxfordshire, was found murdered in a field near his home. The suspect, a sheep-thief named Benjamin Tyler, denied the crime and the evidence was initially found to be circumstantial. It transpired in evidence that Edden's wife had requested Tyler to touch her husband's corpse to test his innocence but he had refused to do so.[39]

The fundamental aspect is not only that the victim's corpse actually bled when touched by the murderer, but the emotional upheaval of those involved and their need to placate the injured soul of the victim as much as their own fears and resentments. The living had to negotiate with the deceased to allow the soul to begin its journey and the corpse to recover its resting place. The living criminal body, through its sympathetic link to the victim, was the agent that enabled the spiritual–medical miracle to occur. There were other related customs concerning corpse-touching, though, that did not concern murderers. At Plymouth in December 1879, during the examination of Mrs. West, murdered by her husband, two servant girls visited their dead mistress and one told the other to touch the body, so 'to prevent dreaming about it, or seeing it again'. For the same reason, in Lincolnshire it was a common tradition to touch a dead body, whether it belonged to a friend or to a stranger, while in the East Riding of Yorkshire, up to the beginning of the twentieth century, it was believed that kissing the corpse ensured never being afraid of the dead. The nineteenth-century folklorist, William Henderson, who thought these traditions had their origins in cruentation, explained that through touching the corpse, the living communicated to the deceased person that they meant no harm, but had only peaceful feelings towards him or her.[40] Touching the corpse was a means to tame it, resolving alleged or actual conflicts between the living and the dead. The issuing of blood in the presence of a murderer suggests, though, that the relationship between blood and soul was crucial to the act of divine trial, and that the other customs were based on a more pragmatic relationship between the living, the dead and afterlife interventions.

Exploring the potency of living criminal bodies in the past throws up a series of interconnected ideas and theories current across the centuries, which were orthodoxy in the early modern period and then became largely restricted to popular cultures by the nineteenth century. Physiognomy, humours, sympathy and contagion all appear as medical factors in how living criminal bodies were thought to express criminality and to have influence on the people around them. Then there were the religious and moral issues of sin, redemption and the fate of the soul. As both a medical and spiritual force, blood appears to be a crucial link between these various medical and religious factors. The witch and the bleeding corpse displayed the notion of crime as both spiritual sickness and physical power. Spiritual crime manifested in the witch as an indispensable, contingent characteristic, able to turn the natural fragility of a human body against society. The body of the witch was clearly potent in life, but there is little in the literature of the early modern period or in the folklore of the nineteenth century that suggests that the corpses of executed or deceased witches were considered to have magical properties. By contrast, cruentation suggests that the common criminal's body, while marked by God or nature, was only actively potent after execution. And this is what we shall now explore.

NOTES

1. Martin Porter, *Windows of the Soul: Physiognomy in European Culture 1470–1780* (Oxford, 2005), p. 12.
2. Katharine Park, 'The Criminal and the Saintly Body: Autopsy and Dissection in Renaissance Italy', *Renaissance Quarterly* 47 (1994) 26. See also, Francesco Paolo de Ceglia, 'Thinking with the Saint: The Miracle of Saint Januarius of Naples and Science in Early Modern Europe', *Early Science and Medicine* 19 (2014) 133–173.
3. Nicole Rafter, 'The Murderous Dutch Fiddler: Criminology, History and the Problem of Phrenology', *Theoretical Criminology* 9 (2005) 76. See also, Nicole Rafter, *The Criminal Brain: Understanding Biological Theories of Crime* (New York, 2008).
4. Anne Harrington, *Medicine, Mind, and the Double Brain: A Study in Nineteenth-Century Thought* (Princeton, 1987), p. 9.
5. Quoted in George Coombe, *A System of Phrenology* (New York, 1842), p. 502.
6. 'Case of William Saville', *The Phrenological Journal*, 17 (1844) 387.

7. See, for example, B.H. Coates, 'Comments on some of the Illustrations derived by Phrenology from Comparative Anatomy—with Reference to a late Review of Dr. Warren's Work on the Nervous System', *The Philadelphia Journal of the Medical and Physical Sciences* 7 (1823) 58–80.

8. *The London and Paris Observer* 13 (1837) 632.

9. David Horn, *The Criminal Body: Lombroso and the Anatomy of Deviance* (New York, 2003), pp. 13–16.

10. Horn, *Criminal Body*, ch. 5.

11. See, for example, Owen Davies, 'Magic in Common and Legal Perspectives', in David J. Collins (ed.), *The Cambridge History of Magic and Witchcraft in the West: From Antiquity to the Present* (Cambridge, 2015), pp. 521–546.

12. See, for example, Sarah Ferber, 'Body of the Witch', in Richard M. Golden (ed.), *Encyclopedia of Witchcraft: The Western Tradition* (Santa Barbara, 2006), Vol. 1, pp. 131–133.

13. Joseph Glanvill, *Saducismus triumphatus, or, Full and plain evidence concerning witches and apparitions* (London, 1681), p. 17.

14. Robert Burton, *The Anatomy of Melancholy* (London, 1621), p. 119.

15. Reginald Scot, *The Discoverie of Witchcraft* (London, 1584), p. 486. Glanvill believed the same. See Glanvill, *Saducismus triumphatus*, p. 23.

16. Jacqueline Van Gent, *Magic, Body and the Self in Eighteenth-Century Sweden* (Leiden, 2009), p. 111.

17. Francesco Maria Guazzo, *Compendium Maleficarum*, trans. E.A. Ashwin (London: 1920), p. 14.

18. William Monter, 'Witchcraft in Geneva, 1537–1662', *Journal of Modern History*, 43 (1971), 179–204; P.G. Maxwell-Stuart, *Satan's Conspiracy: Magic and Witchcraft in Sixteenth-Century Scotland* (East Linton, 2001), pp. 122–124; Barbara Rosen, *Witchcraft in England 1558–1618* (Amherst, 1991), p. 194; Gustav Henningsen, *The Witches' Advocate: Basque Witchcraft and the Spanish Inquisition (1609–1614)* (Reno, 1980), pp. 73–76, 117; Jens Christian V. Johansen, 'Denmark: The Sociology of Accusations', in Bengt Ankarloo and Gustav Henningsen (eds), *Early Modern European Witchcraft: Centres and Peripheries* (Oxford, 1993), pp. 339–366.

19. John Phillips, *The Examination and confession of certaine Wytches at Chensforde* (London, 1566), no pagination.

20. *A Rehearsal both straung and true, of hainous and horrible actes committed by Elizabeth Stile* (London, 1579), no pagination.

21. *A true and just Recorde, of the Information, Examination and Confession of all the Witches, taken at S. Oses in the countie of Essex* (London, 1582), no pagination.

22. Charles Ewen L'Estrange, *Witchcraft and Demonianism* (London, 1933), pp. 275–276; *A true and impartial relation of the informations against three Witches, Temperance Lloyd, Mary Trembles and Susan Edwards* (London, 1682), p. 11; *The full tryals, examination, and condemnation of four notorious witches at the assizes held at Worcester* (London, 1690), p. 6.

23. Pico della Mirandola, *Libro della strega o delle illusioni del demonio* (Bologna, 1524), p. 129; Tommaso Garzoni, *La piazza universale di tutte le professioni el mondo, e nobili et ignobili* (Venice, [1585] 1601), p. 836.

24. Matteo Duni, *Under the Devil's Spell. Witches, Sorcerers, and the Inquisition in Renaissance Italy* (Florence, 2007), p. 119; Marina Romanello (ed.), *La stregoneria in Europa (1450–1650)* (Bologna, 1978), pp. 119–131.

25. Franco Cardini (ed.), *Gostanza, la strega di San Miniato* (Bari, 1989), p. 120; Giuliana Zanelli, *Streghe e Società nell'Emilia Romagna del Cinque-Seicento* (Ravenna, 1992), pp. 69–70.

26. For examples of scratching: *A Rehearsal both straung and true, of hainous and horrible actes committed by Elizabeth Stile, Alias Rockingham, Mother Dutten, Mother Devell, Mother Margaret* (London, 1579); *A detection of damnable driftes, practized by three vvitches arraigned at Chelmifforde in Essex* (London, 1579); Charles L'Estrange Ewen, *Witchcraft in The Star Chamber* (London, 1938), pp. 18–19; Ewen, *Witchcraft and Demonianism*, p. 284; W.J. Hardy (ed.), *Hertford County Records* (Hertford, 1905), Vol. 1, p. 137.

27. Ewen, *Witchcraft and Demonianism*, pp. 170–171, 190–191, 231, 193. See also, I.D., *The Most wonderfull and true storie, of a certaine witch named Alse Gooderige of Stapenhill* (London, 1597), p. 9.

28. J. Raine (ed.), *Depositions from York Castle*, Surtees Society 40 (Durham, 1860), p. 58.

29. Davies, *Witchcraft, Magic and Culture, 1736–1951* (Manchester, 1999), pp. 196–197.

30. *John O'Groat Journal*, 5 September 1856.

31. Owen Davies, *A People Bewitched: Witchcraft and Magic in Nineteenth-Century Somerset* (Bruton, 1999), pp. 125, 134; Davies, *Witchcraft, Magic and Culture*, p. 200.

32. Davies, 'Magic in Common and Legal Perspective', pp. 534–536; *Edinburgh Evening News*, 30 November 1880; *Sheffield Daily Telegraph*, 19 July 1890.

33. Sara M. Butler, *Forensic Medicine and Death Investigation in Medieval England* (New York, 2015), p. 141.

34. Malcolm Gaskill, *Crime and Mentalities in early Modern England* (Cambridge, 2000), pp. 227–231.

35. James VI, *Daemonologie* (London, 1597), p. 80.
36. John Webster, *The Displaying of Supposed Witchcraft* (London, 1677), p. 308.
37. Webster, *The Displaying of Supposed Witchcraft*, pp. 309–310.
38. Gaskill, *Crime and Mentalities*, p. 228. See also, Katherine D. Watson, *Forensic Medicine in Western Society: A History* (London, 2011); Robert P. Brittain, 'Cruentation: In Legal Medicine and in Literature', *Medical History* 9 (1965) 82–88.
39. *Berkshire Chronicle*, 13 October 1832; *Jackson's Oxford Journal*, 13 March 1830. Tyler and his accomplice were subsequently hanged at Aylesbury.
40. *Reports and Transactions of the Devonshire Association for the Advancement of Science Literature, and Art* 12 (1880), 103; Mrs. Gutch, *County Folklore: East Riding of Yorkshire* (London, 1911), p. 135; William Henderson, *Notes on the Folk-Lore of the Northern Counties of England and the Borders* (London, 1879), p. 57.

The Corpse Gives Life

Abstract This chapter begins by looking at the trade in human fat into the nineteenth century, and how the control over its availability switched from the executioner to the anatomy schools. This, and other developments, led to the decline of the executioner–healer on the Continent. The similar trade in human skin for macabre mementos and magic is explored. The chapter then considers the history of the healing touch of the hanged man's hand in England, and the rise of blood-drinking at beheadings in nineteenth-century Germany and Scandinavia.

Keywords Human fat · Blood-drinking · Human skin · Hanged man's hand · Healing touch

The French journalist, Félix Pyat, observed humorously in 1841 that 'the executioner is a bit of a doctor, just as the doctor is a bit of an executioner'.[1] However, by this time the role of the executioner-healer was vanishing. As Kathy Stuart has discussed with regard to Germany, during the early modern period the medical side of the executioners' profession had given the dishonourable trade an important route to social mobility. At least nine executioners' sons matriculated from the University of Ingolstadt, Bavaria, between 1680 and 1770, for instance. However, restrictions regarding executioner medicine, and the abolition of torture across western and central Europe during the late eighteenth century,

© The Author(s) 2017
O. Davies and F. Matteoni, *Executing Magic in the Modern Era*,
Palgrave Historical Studies in the Criminal Corpse and its Afterlife,
DOI 10.1007/978-3-319-59519-1_3

reduced the financial rewards of the trade, thereby limiting avenues for social advancement.[2] In Spain, for example, a law of 1793 suppressed all the perquisites of the executioner, ending the sale of criminal corpse parts for medicine.[3]

The medical respectability of the executioner was, nevertheless, still evident during the early years of the nineteenth century, though in Britain there was no such tradition of physician-hangmen. The executioner of Lyon, named Chrétien, had a considerable reputation as a healer in the region, particularly for a secret unguent he sold for the treatment of rheumatism.[4] In Norway, Erik Petersen (1766–1835), a chairmaker by trade, was appointed public executioner for Trondheim in 1796, while also serving as an assistant in the town's hospital and, in 1808, as an army surgeon during the war with Sweden. As we saw in the Introduction, at around the same time, the Bornholm executioner, Caspar Frederik Dirks, held the post in large part because of his medical credentials. When he threatened to resign in 1791, an army surgeon put in an application to replace him.[5] However, the early decades saw the growing success of the orthodox medical profession in securing a monopoly on medical practice through the suppression of irregular, unlicensed healers, and the imposition of qualification requirements for public health positions. Therefore when, in 1818, Erik Petersen applied for a medical licence, he was turned down, despite his extensive experience in the field.

In his seminal study of professional and popular medicine in France between 1770 and 1830, Michael Ramsey notes how some executioner-healers tried adapting to the new realities regarding public health legislation and state-controlled health provision, and sought formal medical qualifications—with some success. During the Revolution, two executioner-bonesetters in Vannes and one in Nîmes took out patents as health officers. In 1820, the executioner for the department of Yonne pursued examinations to be a health officer. When, in 1819, an experienced executioner and bonesetter named Caron, formerly executioner in the Gironde, took on the role of executioner in Bordeaux, a local physician made a formal complaint. An investigation revealed that Caron had received a legitimate health officer's diploma in anticipation of becoming the executioner for Versailles, but subsequently fraudulently obtained a Paris university medical diploma.[6] Reforms after 1830 further undermined the French executioner's position, though. In 1832, a law regarding the replacement of aged *bourreaux* began the process of reducing the number of executioners and also their assistants.

During the 1840s, further changes reduced the pay of some executioners in a period when there were relatively few executions. For some, their subsidiary medical practice became more important, or, ultimately, their sole source of income. In 1852, for example, the ex-*bourreau* of Chalon-sur-Saône, Etienne-Théodore Cané, announced in the local press that he was a 'surgeon-dentist technician' offering orthopaedic bandages for hernias and supports for other deformities of the body. However, the French medical profession was hawk-eyed in its crusade to eradicate unlicensed medical practitioners. The general practitioners of Agen, for instance, made a formal complaint against the local *bourreau*, Jean-Baptiste Champin, who had a large signboard on his house advertising his services as a '*bandagiste patenté*' (licensed bandager), hernia sufferers being his principal clientele.[7] Henri-Clément Sanson, Paris executioner between 1840 and 1847, recalled how his father Henri, who operated the Paris guillotine from 1795 to 1840, had relied heavily on the medical formulas and recipes held by the Sanson family. His healing successes were so well known, he claimed, that eminent surgeons did not disdain from sending on patients to be healed. However, as Henri-Clément noted with frustration, in his own times, 'the rigour with which they were demanding that practitioners had diplomas removed from us a large part of our clientele'.[8]

THE CRIMINAL CORPSE IN PIECES

As we have just seen, the executioner's role as healer staggered on into the mid-nineteenth century, but what of the trade in body parts, which used to be a central aspect of their ancillary medical and magical business? Human fat had been one of the most important substances in those states where the executioner had the right to dissect his victims. During the seventeenth century, in particular, it was traded and employed for medical purposes in Germany, Italy, France and Spain. The commerce in human fat was mostly medical, though there was some discussion about the magical (or diabolical) power of the meat of executed criminals, particularly in love magic rituals.[9] It was deemed efficacious against contusions, broken bones, the scars of smallpox and wounds. In the early eighteenth century, one Italian charlatan sold an unguent called *Balsamo del sole* (Balm of the Sun) to treat cold and damp humours, attributed to the alchemist Giuseppe Borri (1627–1695), which purportedly contained human fat. A Roman charlatan named Lorenzo Sabatini sought permission to sell his expensive 'Balm of Human Fat' from the corpses of healthy men who had been executed or recently killed in accidents. It was for the cure of pneumonia and gout.[10] However, the executioner's right to dissect criminal corpses was largely a thing of the past by the mid-eighteenth century. In 1742, the Augsburg executioner, Johann George Trenkler, was ordered, for instance, to hand over all criminal cadavers to the hospital for dissection by medical students. Five years later, he petitioned to have the right to dissect for the good of public health. It was the best source of poor sinners' fat, and he claimed, 'The whole town knows that by mixing a salve with human fat .... I have cured several patients of their nerve gout.'[11] He was unsuccessful. In 1799, the Justice Committee in Turin, Italy, recommended compensating the city executioner the sum of 24 *soldi* (silver coins) for the income lost once he was forbidden to sell human fat.[12] However, as Kathy Stuart has observed, 'the use of human fat was not particular to executioner medicine; executioners simply had privileged access to it'.[13]

During the second half of the eighteenth century, the influential Italian anatomy professor, Domenico Cotugno (1736–1822), still listed human fat as a remedy against gout or nervous sciatica, yet he affirmed that it was no more potent than other fats. What made the difference was its alleged scarcity and the power that people attributed to it.[14] Prices were certainly high by the mid-eighteenth century. In 1761, the well-known Madrid apothecary, José Hortega, was selling it for the considerable sum of 160 *reales de vellón* per pound. The trade in pre-Revolutionary France was also a highly competitive business. One pharmacist advertised that he had fought the executioners and the apothecaries to be able to sell his human fat, which he boasted was better than that of the hangman because of the special seasoning he applied to stop it going rancid.[15] Seventeenth- and early eighteenth-century medical texts that mentioned human fat circulated long after the substance was no longer readily available. Popular herbals and household guides, such as *l'Agriculture et maison rustique*, which went through numerous editions throughout the seventeenth century, and was no doubt consulted in homes for decades after, contained several references to using human fat in home remedies. It recommended mixing it with the herb *elatine* (waterwort) to ease the pain of gout, and noted it was also 'miraculous' in drawing out harquebus shot from the body.[16]

It is no surprise, then, that the commerce in human fat did not cease completely with the ending of the executioner's perquisite. While demand remained, so money was still to be made. If the executioner could no longer supply it, the new masters of the criminal corpse, the medical profession, were not beyond illegal trade in the substance, as was revealed by a sensational case in France. During the early nineteenth century, Paris medical students and their dissection theatre assistants, the *garçons de l'amphithéâtre*, were found to be systematically draining and collecting human fat from the corpses they dissected and selling it illegally for a variety of purposes. The principal buyers were enamellers, and other craftsmen who used blow torches, because it was thought human fat produced a steadier, stronger flame compared with the legal supply of dog and horse fat. It was also sold as axle grease, and purchased by unorthodox healers and druggists as part of their pharmacopoeia. This public hygiene scandal was investigated by the authorities in 1813, though to avoid raising public fears reporting restrictions were enforced, so the press did not get a whiff of the story at the time. Those culpable were sent to prison for 6 months. Two thousand litres of human fat were discovered in the care of a man at one of the medical schools, 400 kg in the possession of another, and smaller amounts in other hands at the dissection amphitheatre.[17] The cholera expert, Charles Londe (1795–1862), recalled that during his student days in Paris a restaurant owner, who served large numbers of students, was sent to prison for having used human fat obtained from the *garçons de l'amphithéâtre* in his cooking.[18]

It is unlikely much of this Parisian fat made it into the distant French provinces, where demand also continued. In Lyon, after the death of the executioner Chrétien, people continued to resort to the city's pharmacies, requesting in vain '*la graisse de Chrétien*'. In his 1869 book on natural and artificial body fats, the Montpelier physician L.-H. De Martin recalled the days when the *bourreaux* sold the '*graisse de pendu*' to the credulous, lamenting 'if hanging has been abolished, credulity, alas! has not been'. A trade clearly remained for longer. A pharmacist from Montbrison in the Haute Loire wrote to the professional *Journal de chimie médicale* in 1860 to complain about the unfair competition he and his fellows faced from the church, itinerant charlatans and grocers. He wrote of one grocery shop he had visited that had fairly recently sold what was claimed to be human fat as part of its pharmaceutical products.[19] One wonders about the real source and nature of the fat being sold in the Montbrison pharmacy. There had, no doubt, all along been a lively trade in passing off pig and goose fat as human.

Other parts of the criminal corpse were desired more for their magical properties. The Franciscan, Bernardino of Siena (1380–1444), who preached against popular magic, noted the belief that to cure toothache one could touch the bad tooth with the tooth of a hanged man. It was recorded in late seventeenth-century France that to wear a wig made from the hair of hanged man soaked in the blood of a hoopoe would render one invisible.[20] In early modern Germany, it was believed that the finger of an executed criminal brought luck to a household or business, and when placed in the stables, the horses would thrive. In the early sixteenth century, one Hans Moller von Dippertswald was prosecuted for fraud for selling the thumbs of executed thieves.[21] In early modern Spanish inquisition records, we find people who covertly sought to steal for magical purposes parts of the criminal corpse as it was hanging or was displayed in public places. In Spain, it was forbidden to remove the hanging corpse until nightfall, so there was a period when raids could be made under darkness, depending on how quick the hangman was in taking down the corpse. In 1586, Ana de Yuso confessed how she and an acquaintance named Geronima 'had gone one night to a hanging and the said Geronima had asked a man who went with them to cut her a piece of the rope or a finger from the body and, when he had unsheathed his sword to do so, another man arrived and prevented him.' In another case, it was testified that the 'defendant was boasting that she had been to the scaffold of this city and had taken the heart from a hanged man ... and that the defendant was in the company of other women at the scaffold one morning and was unable to cut the hand from a dead man who was hanging there because people had come past'.[22] As these suggest, stories and boasts of obtaining body parts probably far outnumbered the actual taking of body parts without the executioner's consent. And, of course, the Spanish switch to the garrotte in the eighteenth century ended such opportunities anyway.

In states where criminal corpses continued to undergo public postmortem punishment in the modern era, there were still opportunities for popular access to pieces of corpse. The Pomeranian High Court expressed concern about the problem in 1811: 'The general superstition of the common mob, that the possession of a limb of an executed malefactor or a piece of his clothing brings good luck, has led to frequent misappropriation of such items from these places.' It had received two recent reports of such activity. A journeyman cobbler had taken a bone from a body left on the wheel in Pollnow. It was reported from Saxony, in 1823, that within a week all the fingers, toes and clothes of an exposed criminal corpse at Schneeberg had been removed for magical and medical purposes.[23] There are some examples in which the body pieces and the identity of the criminal were expressly related: something of the former notoriety of the criminal remained sealed and transmissible inside them. This is the case with people convicted for political crimes—revolutionaries and rebels who became heroic figures for the crowd. The story of the Tyrolean rebel, Andreas Hofer, executed in Munich in 1810, is a good example. Apparently, some soldiers attempted to get hold of a limb from his corpse as a protective talisman, but they were caught and punished.[24]

There is little evidence of people in England raiding gibbets for corpse pieces during the eighteenth or early nineteenth century. Only a handful of gibbetings took place after 1800, and the authorities went to considerable lengths to ensure there was no cutting down or tampering of gibbeted corpses for whatever purpose. The gibbet posts were often more than 10 m high and spikes were sometimes fitted around the base.[25] However, up until the Anatomy Act of 1832, the policy of dissecting male criminal corpses provided potentially new opportunities for obtaining body parts for private purposes. There is no evidence in England of human fat being collected from the dissection room and sold as in Paris, though. As Elizabeth Hurren observes, apart from the skeleton and tanned skin, 'most human remains simply disappeared down the drain, as criminal flesh disintegrated into dusty sweepings'.[26] While skeletons ended up in museums and private collections, it was the skin that had the most diverse afterlife. Belts of tanned human skin, for instance, had long been used in medicine as a cure for labour pains. In the 1680s, the German physician, Johann Jacobi Waldschmidt, noted that it came highly recommended for this purpose, and a Danish pharmacopoeia of around the same date quoted the current price as 16 marks a piece.[27] The trade in magic belts accordingly declined as executioners ceased to have control over the criminal corpse.

One of the most interesting biographies of a criminal corpse skin is that of the English murderer, Mary Bateman. She was a Leeds cunning woman found guilty of deliberately poisoning one of her clients in 1809 and was hanged by William Curry. Her corpse was publicly dissected, with £80 14s being raised from entrance charges. At least one body part was circulated, with the tip of Mary Bateman's tongue finding its way into the curio collection of the governor of Ripon prison.[28]

More money changed hands for pieces of her tanned skin. Two books were commissioned to be bound in it, Sir John Cheeke's *Hurt of Sedition: How Grievous it is to a Common Welth* (1569), and Richard Braithwaite's *Arcadian Princess* (1635), which used to be in the library of Mexborough House, Yorkshire, but mysteriously went missing when the library was being catalogued for sale in the mid-nineteenth century.[29] There was a long European tradition of covering books, often medical texts, in flayed human skin, a practice known as anthropodermic bibliopegy. In early nineteenth-century Britain, before the Anatomy Act of 1832, there was a vogue for covering published accounts of a murderer's trial with the skin from his or her corpse.[30] This was the fate of William Corder's epidermis. Convicted for the sensational Red Barn murder case, he was executed at Bury St. Edmunds in 1828. 'The binding of this book is the skin of the murderer, William Corder,' states a note written in the copy, 'taken from his body and tanned by myself in the year 1828. George Creed, surgeon to the Suffolk Hospital.'[31]

Small strips of Bateman's skin were also tanned and sold as curios. The York surgeon, Richard Hey, wrote to the *Medical Times* in 1856 to say that his grandfather had given several anatomical lectures using Bateman's body at Leeds infirmary, and his 'eccentric' father had kept a piece of her tanned skin.[32] A farmer named John Andrew of Birstwith, North Yorkshire, also possessed one such piece in the 1850s. Later in the century, strips preserved in jars were still circulating in Yorkshire.[33] Pieces of her skin were not only desired as curios, they also apparently accrued a reputation for having the power to ward off evil spirits. However, this is the only English example of criminal corpse skin being used for such magical purposes, and one suspects that Mary Bateman's reputation as a cunning woman was key to why her skin might have accrued such a reputation, rather than it being a criminal's skin per se.[34]

In late nineteenth century France, a skin scandal broke that echoed the fat scandal of 1813, revealing once again how the old perquisites of the executioner had passed on to the new 'honourable' custodians of the criminal corpse. In 1883, Michel Campi, who become known as 'the mysterious assassin of the Rue du Regard', confessed to cutting the throats of a retired Parisian advocate named Ducros and his elderly sister. Campi's past was surrounded in mystery, which he was keen to promote during the investigation, but he was executed the following year after a sensational trial.[35] His body was taken for dissection, and his skeleton preserved and put on public display at the Broca Museum of the

Laboratoire d'Anthropologie. The fate of his skin became something of a *cause célèbre* over the next few years. The national newspaper, *Le Figaro*, reported in a scoop, in November 1884, that the skin of Campi's right side and arm had been taken and tanned under the instructions of one Flandinette from the *Laboratoire*, who wanted to cover his account of the Campi case in the murderer's skin.[36]

A few years later, another skin scandal broke in the press regarding the corpse of the murderer Henri Pranzini, executed at the Roquette Prison, Paris, in August 1887. Some skin had been taken from his chest, tanned, and used to cover several wallets that were given to senior officials. A satirical verse on the scandal ran:

> If you want to hold on to your skeleton,
>
> Don't do like Pranzini.
>
> Die in your bed
>
> Rather than snuff it at the Roquette.
>
> Your body will not be cut up,
>
> In pieces for the *Sûreté*.[37]

Under pressure from the newspapers, the wallets were destroyed. One of those who had been gifted one, Marie-François Goron, Head of the *Sûreté*, wrote about the affair in his memoires, and expressed the view that the press had blown the matter out of all proportion. Despite Goron's best efforts, the *garçon de l'amphithéâtre* involved in the transaction was dismissed by the Faculty for abetting in the taking of the skin from the dissection room. The poor man died shortly after—'truly the posthumous victim' of Pranzini, wrote Goron.

And what of the fate of Campi's skin that had been tanned for unknown purposes? In 1887, it was stated that his skin was now 'worth its weight in gold for some people'.[38] And it was not just his skin that was desired. A journalist visiting the Broca Museum in 1886 reported that some finger bones and the thumb of the right hand had been stolen. When asked the reason why, the staff said they thought it was probably students seeking curiosities.[39] Maybe or maybe not. Were the anatomy museums a new location, replacing the execution site, for the pilfering of criminal body parts for magical and medical purposes? As to Campi's skin, well, the Pitt Rivers Museum in England possesses a 'policeman's lucky amulet originally acquired in Paris' in 1889, which consists of a pierced *sou* attached to a small piece of tanned skin, supposedly Campi's, and a piece of cord purportedly from a hangman's rope.[40]

THE FRESH BODY

The trade in criminal body parts for medicine and magic was dependent on either the post-execution business of medical dissection or the illegal in situ theft of body parts when criminal corpses were left on display and unguarded long after execution. However, two other healing traditions that continued into the mid-nineteenth century were dependent on gaining access to the corpse minutes after the execution. One concerned the drinking of corpse blood, which will be discussed shortly, and the other was the touch of the hanged man's hand.

In 1854, the famed French novelist and campaigner for the abolition of capital punishment, Victor Hugo, wrote an account of the hanging of the murderer, John Tapner, the last person executed on Guernsey. Hugo was living on the neighbouring island of Jersey at the time and followed the case closely, writing a letter of protest to the British foreign secretary, Lord Palmerston. Tapner's execution was meant to be held in private, but some 200 ticket-holders were present. Although he did not attend, Hugo described the scene as reported to him: 'Tapner dead the law satisfied. It is now the turn of the superstitious; they never failed to come to the rendezvous which the gallows gives them. Epileptics came, and could not be prevented from seizing the convulsive hand of the dead man and passing it frantically over their faces.'[41] This is the last known case of people having access to the hanged man's hand for a cure in the British Isles, the last permitted example on mainland Britain being at a Warwick execution in 1845. A journalist described the latter instance:

> The body remained suspended the usual interval of time, during which was enacted one of those disgusting scenes of vulgar superstition, which, in these days of boasted enlightenment, it was as extraordinary as it was revolting to behold. We allude to several females being allowed to ascend the scaffold, to have their wenned-necks rubbed by the still warm hand of the malefactor.[42]

The authors have found over two dozen cases of people requesting to be touched by the hand of a freshly hanged corpse at executions between the mid-eighteenth and mid-nineteenth centuries, mostly to cure swellings on the neck or head caused by goitre (enlarged thyroid gland), scrofula (infection of the lymph nodes, often associated with tuberculosis) and wens (sebaceous cysts). The hanged man's touch was primarily practised in southern England, though the notion that the hand of the freshly deceased, and of suicides in particular, could similarly cure was widespread. Related traditions also appear on the Continent. In the mid-nineteenth century, French Basque people continued to believe that the executioner of Pau could heal goitre by touching the patient's neck.[43]

The reason why the touch of the hanged man's hand was thought to be efficacious is unclear, and multiple explanations may have circulated in popular culture. The rules of magical transference may have been at work, with the illness being passed from the living patient to the still-warm hand of the hanging criminal, whose soul, not yet departed from his quivering body, would then take the ailment into the afterlife. Maybe there was no such spiritual connection, and it was merely an act of simple sympathetic magic. Once touched by the hand, the swelling would reduce as the buried or gibbeted corpse decayed and disintegrated. There was a similar widespread notion that warts would disappear by touching them with a piece of meat and then burying it. Why the executed criminal, as distinct from any other corpse? Perhaps it was considered extra-potent due to the redemptive quality of the criminal's last confessional moments on the scaffold. Or maybe the explanation is purely a pragmatic one, with executions offering the easiest access to a corpse so quickly after death. Knowing in advance the time and place of an execution enabled the sick to plan their intended appointment with the freshly deceased.

Access to the hanged man's hand was carefully controlled by the executioner and became one of his perquisites. Just before the corpse of James Morgan was cut down at a hanging in Maidstone in 1819, for instance, the hangman enabled a young woman to stand on the waiting coffin in order to be able to reach up and have the hand of the hanging corpse passed over a swelling on her throat.[44] The hangman would usually take the hand of the corpse and administer the rubbings to the patients as they stood or sat on the scaffold. The condemned were aware that their bodies might be exploited in this way minutes after their death. In 1815, while on the Newgate scaffold awaiting his hanging, the forger, John Binstead, requested of the presiding clergyman that his hands would not be made available to those seeking a cure for wens.[45] The process accrued magical elements, with the stroke sometimes being given nine times, three and nine being important numbers in folk magical ritual. There is clear evidence for the importance of contrasexual charming as well: in other words, that the cure would only work if a woman touched a male hand and vice versa. When Ann Norris and Samuel Hayward were hanged at Newgate on 27 November 1821, the *Morning Post* reported that 'Several men rubbed their necks and faces with the hand of the unfortunate female—while as many women went through the same ridiculous and indecent charm with those of the wretch Hayward.'[46]

The increasingly theatrical and ritualised nature of hangings towards the end of the eighteenth century, with the scaffold platform placed high above and at some distance from the curious crowds, gave the whole procedure of obtaining the stroke a formalised and official status. The clampdown on the practice began in the early nineteenth century, with the newspapers repeatedly requesting the authorities do something to suppress what was described as a barbaric superstition. However, it proved a remarkably tenacious practice, considering the ease with which the authorities could have stopped it, as they controlled every aspect of the execution procedure. The hangmen were reluctant to give up a valuable extra income, and the sheriffs, who were responsible for hiring and paying executioners, were reluctant to antagonise the small pool of hangmen. The eventual suppression of the practice was apparently primarily due to the growing influence of prison governors, as prisons became the sole location for executions during the nineteenth century. The authors have written at length about the hanged man's hand tradition in an article free to the public,[47] so let us pass on to the matter of obtaining fresh blood at continental beheadings, which raises similar issues regarding state policy, spectacle and popular medical and magical tradition.

To the modern imagination, the drinking of a criminal's blood immediately after his execution might appear as some barbaric 'medieval' superstition or some grotesque story of diabolical ritual from the torture chambers of the witch trials, but how widely it was practised from the sixteenth to the early eighteenth century is unclear. Most of our recorded accounts concern the nineteenth century: some eighteen or so cases of the practice from late-eighteenth- and nineteenth-century Germany, nine or so cases from Sweden, four from Denmark and three from Norway.[48] The last recorded instances across northern Europe were in the 1860s, while the last unsuccessful request to get blood from an executed criminal to heal epilepsy was made by a woman at Freiberg in 1908.[49] As with the hanged man's touch, there were undoubtedly numerous other instances that we have not come across or which went unreported. The prevalence of the practice depended, of course, on execution techniques, and, as we saw in the Introduction, Sweden only switched from hanging to beheading in the early nineteenth century. Norway, likewise, abolished hanging in 1815 and turned to the chopping block, so in terms of popular execution magic and medicine, one door closed and other opportunities opened up as the blood ran freely. As with the hanged man's hand, what seems to be a macabre survival of superstition was actually a modern, state-enabled, popular medical tradition.

The pattern of demand for the practice also needs to be considered in the context of the significant reduction in executions in Scandinavia during the period, as sentences were increasingly commuted and imprisonment became the norm for capital offences. In Denmark, there were only 42 beheadings between 1841 and 1892, while in Sweden over 600 executions took place between 1800 and 1864, mostly concentrated in the early decades, with only 15 between 1866 and 1910.[50] Thus, when a butcher named Marcusson and his accomplice in murder, Bottilla Nilsdottir, were due to be executed in Ystad, Sweden, in February 1851, a large number of people assembled to obtain their blood. It was noted, after all, that there had been no execution in Sweden for 8 years, and no execution in that province for 60 years. The pent-up demand was clearly huge.[51]

It is not clear to what extent blood-drinking was controlled by the executioner as one of his perquisites. There seems to have been a free-for-all on some occasions. Take this account given by the American travel writer, John Ross Browne, of a beheading in Hanau, near Frankfurt am Main, during his visit to the country in 1861:

> Standing near the scaffold, in close proximity to the criminal, within the guard of soldiers, were six or eight men from the mass of the people, said to be afflicted with epilepsy. The moment the head was off these men rushed to the body with tumblers in their hands, caught the blood as it spouted smoking warm from the trunk, and drank it down with frantic eagerness! Their hands, faces, and breasts were covered with the crimson flood that ebbed from the heaving corpse. One man, too late to catch the blood as it spurted from the neck, took hold of the body by the shoulders,

inclined it over in a horizontal position, poured out his tumbler full from
the gory trunk, and drank it in a wild frenzy of joy![52]

In most cases, though, one gets a clear sense of the procedure being
curated by the executioner. At the beheading of Carsten Hinrich Hinz,
at Tonning in April 1844, the executioner allowed the epileptic son of a
farmer from Oldenburg to drink some of the criminal's blood. Likewise,
at a beheading in Hanover in 1857, it was reported that a number of
epileptics came forward and the executioner 'readily gave it to them'.[53]
From Sweden, we have one example in January 1833, when, in the parish of Hällefors in Västmanland, people came to acquire blood from
the executioner at the beheading of the murderer, Olof Liljeblad.[54] A
degree of formal process was exhibited at a public execution in Mainz,
in 1802, where the executioner's assistants caught the blood in a beaker
and then some of the crowd were allowed to come forward to drink it.
Again, in Berlin in 1864, the executioner's assistants distributed white
handkerchiefs dipped in the victim's blood, selling them for two *thalers*
a piece.[55] By custom, it was not only the authorities who dictated access
to the criminal's blood: in Scandinavia, the sick also occasionally sought
the permission of those awaiting beheading. The day before an execution was due to take place in Dahlby in Scania, Sweden, numerous sick
people visited the jail and obtained from the criminal permission to drink
his blood—'a needful proceeding to render the loathsome draft efficacious'.[56] In Assens, Denmark, in August 1856, two girls aged 15 and
17 collected and drank the warm blood of Peder Olsen, who had murdered a young woman. When a magistrate questioned them regarding
their behaviour, they testified that there was nothing wrong: they had
received the signed permission of the condemned to drink his blood.
Olsen's sentence had stated that his head was to be displayed on a pike,
though this was dropped on appeal. Olsen apparently found the state's
use of his body more objectionable than the people's use of his blood.[57]

Pressure from the press and commentators, who decried such public
exhibitions of 'credulity' and 'ignorance', led to the suppression of the
practice during the mid-nineteenth century. A case of blood-drinking at
an execution in Örebro, Sweden, in 1846, led the newspaper *Aftonbladet*
to describe it as 'foul' superstition, for example.[58]

As with the hanged man's hand, there were also law and order concerns about allowing the practice. The task of suppressing the custom
was not easy. In the 1851 Ystad case mentioned above, the authorities

were determined not to let the crowd, armed with their cups, pots, bowls and pans, get access to the blood. Soldiers ringed the execution platform, and at the moment of the beheading, they drove back the surging crowd with their rifle butts. It was reported that the struggle left some 200 people badly injured and many others bruised. By the time some of the blood-seekers had broken through the cordon, the corpses and heads had already been removed and carried away under cavalry escort. The authorities even went to the effort of removing the earth and turf splattered by the blood.

The drinking of blood to cure epilepsy, like so many folk medical traditions, including the aforementioned stroking of a dead man's hand, dates back to antiquity. In his *Natural History*, Pliny the Elder (d. 79 AD) described how epileptics were drawn to gladiatorial spectacles in the arenas—just as they were to the scaffold in the nineteenth century. 'While the crowd looks on,' he observed with disgust, 'epileptics drink the blood of gladiators, a thing horrible to see ... they think it most efficacious to suck as it foams warm from the man himself.'[59] According to ancient Galenic medicine, which held sway through much of the early modern period, blood was one of the four humours and so sometimes more or less blood was required to restore health.[60] As one late seventeenth-century treatise explained, 'good blood is that which is temperate in the first Degree, not too thin not too thick; not sharp nor biting, not bitter, not salt, not sour'.[61] According to Galenic medicine, an excess or plethora of blood, or corrupted blood, was a health problem, leading to the widespread practice of blood-letting. However, blood cures were not dependent on humoral theory. That arch-critic of Galenic medicine, Paracelsus, also recommended blood from a decapitated man drunk at certain astrologically propitious moments as one of several chemical cures for epilepsy, including the ingestion of ground mistletoe or peony seeds.[62]

Some eighteenth- and nineteenth-century German and Scandinavian descriptions of execution blood-drinking show that the ingestion of the blood alone was considered insufficient to effect a cure. Having imbibed from the blood spurting from the corpse's neck, the epileptics had to run fast, often spurred on or supported by family, friends, or bystanders, until they fell down senseless. This therapeutic method was reported in Dresden in June 1755 and in Hanover in 1812. In Zwickau in December 1823, several people, mostly children, drank from a pot full of the executed criminal's blood and were then whipped and obliged to run across a field.[63] Two cases from Denmark demonstrate a similar procedure. The first took place at Skælskør in 1823, and was famously described by the storyteller, Hans Christian Andersen. The second occurred in 1834, at the execution in Hjordkær, North Schleswig, of H.S. Fallesen for murdering his bride. Three epileptics were allowed to drink his blood and were then made to run.[64] In Dahlby, in Scania, no fewer than 70 individuals, suffering from different ailments, came to drink the criminal's blood. Those who managed were then 'seized hold of by two of the bystanders, by whom they are run backwards for some little distance, when the cure is supposed to be effected'.[65] The cure could end tragically, as happened in Stralsund in 1814, when, after having been bound between two horses 'and pulled away at a breakneck gallop', the patient collapsed and died.[66]

A folklore record of the practice from Västergötland, Sweden, explained in scientific fashion that, after having gulped down the blood, people had to run to let the fluid circulate in their own blood and so be better absorbed into it.[67] This notion echoed theories of the blood contained in early modern medical literature. The seventeenth-century Paracelsian chemist, Johann Schröder, wrote that, when drunk, blood caused a 'violent motion' of the patient's own blood and a copious sweating. The doctor explained that it had to be given cautiously to epileptics, because 'it not only brings a Truculency to the takers, but also the Epilepsie'.[68] The theories about epilepsy of another physician, Georg Ernst Stahl (1659–1734), help us to understand this ambivalence. Discussing epilepsy as the result of the soul's attempt to expel plethoric humours or intrusive substances, he listed amenorrhoea in women among the causes of the sickness, but also an accumulation of blood in some organs.[69] Thus, good blood could cleanse the unbalanced body, but an excessive dose could be equally damaging.

There is some evidence during the early nineteenth century that epileptics continued to be referred to the scaffold by physicians. The old humoral and Paracelsian medical theories had been largely abandoned in orthodox medicine, but new or updated notions of psychological and emotional influence on the body seemed to explain the apparent beneficial effect of execution blood. The influential eighteenth-century medical writer, William Cullen, referenced Pliny's account of the gladiators when trying to explain the apparent success of blood-drinking. 'As the operation of horror is, in many respects, analogous to that of terror,' he explained, 'several seeming superstitious remedies have been employed for the cure of epilepsy; and, if they have ever been successful, I think it must be imputed to the horror they had inspired.' The same psychological efficacy of spectacle and terror was attributed to the hanged man's hand cure.[70] This interpretive shift is evident from a case at Stockhausen, Hanover, in 1843, where several epileptics gathered at an execution with their mugs. They were initially denied access on the advice of a local medical man, who declared the cure ineffectual, but they subsequently secured a certificate from medical professors at Göttingen University, to the effect that drinking the blood would have a beneficial 'psychological' effect.[71] John Ross Browne discussed the practice of execution blood-drinking with 'intelligent' German acquaintances, 'who certainly do not share in the superstition, but who still maintain that there may be some reason in it'. The argument they put forward was that, epilepsy being a nervous disease, the shock of witnessing the execution and the act of drinking from the corpse could jolt the nervous system in a beneficial way.[72]

We need to consider whether and why an executed criminal's blood was considered more potent than other people's blood. Maybe beheadings were merely the most convenient legal means of obtaining fresh blood. It is not exactly easy, after all, to obtain copious quantities of warm fresh human blood without resorting to murder. In countries where beheading was not practised, we find epileptics resorting to drinking their own blood. In 1863, for example, a doctor in north-western Scotland recorded that he had seen several epileptics drinking small quantities of their own blood to cure themselves.[73] Animals' blood could also act as a substitute. As with the hanged man's hand, there was, then, undoubtedly a pragmatic element to resorting to the scaffold corpse. However, there were other possible elements at play for why an execution corpse was sought. Galenic humoral theory and Paracelsian notions, both of which filtered into popular medical conceptions, allowed that the heightened passions and emotions of condemned criminals could change the quality of the blood coursing through their bodies. The Danish examples of requesting permission from the condemned criminal provide further pause for thought. Were they merely a demonstration of decency and respect? Or were the applicants fearful of retribution from the future restless, outraged spirit of the condemned? Was the blood thought to be filled with the grace of repentance and purged sin?

Some blood-drinking operated through ritual. In other words, the curative potency of the criminal blood was activated or enhanced by magical or religious observance. In June 1861, a poor woman suffering from epilepsy obtained permission to go to Trogen, Switzerland, and attend an execution in order to try the blood remedy. She was advised to drink it warm in three sips, while repeating the holy trinity. Unfortunately, once at the scaffold, a new epileptic attack prevented her from imbibing the fresh blood.[74] It is clear, furthermore, that the executioner's role was not always merely that of a custodian. With the hanged man's hand, the efficacy of the cure sometimes depended on it being performed by the executioner, hence the notion, in France and Belgium, that the executioner's stroke using his own hand was potent in itself. In some Scandinavian cases, there is evidence that cunning folk also played a mediation role. At the execution of Carl Petersson in Skarpåsen, southern Sweden, in January 1847, a soldier named Olaf Magnusson Bly apparently witnessed two wise women, amongst others, gathering around the corpse to collect the healing blood. When the fluid ran out, he added, the women milked the carotid to get extra drops.[75] At an execution in 1863, carried out by Norway's penultimate state executioner, Samson Isberg, in Etterstad, near Oslo, a wise woman came with three pots in order to fill them with blood to use against epilepsy.[76]

However, no agency was required in other overtly magical uses of execution blood that were not concerned with healing. According to German folklore, if bakers, innkeepers and merchants dipped a piece of linen soaked in the blood of a beheaded criminal into their dough, beer or brandy barrels, they would attract a greater number of customers.[77] At the beheading of the murderer, Treiber, in Munich in September 1852, as well as the usual epileptics, some people sought the blood as they thought it would enhance their chance of drawing lucky numbers in the lottery. Two years later, at an execution in the German town of Adelsheim, people gathered to get some blood because it was thought to act against sickness and evil forces, keeping witches away from houses and stables, and protecting against lightning.[78]

Thus, there clearly was something more than just convenience in the resort to execution blood. Any old blood would not do in magical terms. This leads to the consideration of whether further potency accrued to specific executed people. In other words, did unusual notoriety or celebrity further enhance the power of the blood? When the national revolutionary student, Karl Ludwig Sand, was executed in Mannheim in 1820, people were reported 'to have stormed at the scaffold, soaked up his blood with kerchiefs, broken up the stool on which he had sat, and distributed the pieces'. Evans has argued that the episode might be either 'a political adaptation of the tradition', or simply just another case of medical employment of blood.[79] Nevertheless, it is undeniable that individual identities could exercise a certain charm. An account from Catholic southern Europe, where cases of blood-drinking were rarely attested, seems to confirm the value of blood when taken from particularly respected executed people. The source is Italian, but it deals with Catalan events. In May 1894, in Barcelona, after the execution of six anarchists, who were garrotted, a woman plunged her handkerchief in the blood dripping from the corpses and then she carefully wrapped it in a newspaper, as if it were a special talisman. Then she invited one of the journalists to follow her example.[80]

There has been a tendency to assume that criminal corpse medicine and magic largely ended as the early modern period drew to a close, thanks to the 'usual' Enlightenment mix of popular education, civilising campaigns, religious rationalism, judicial policy and scientific advancements. As recent studies of magic and witchcraft have shown, though, this is a narrative dependent on a now-debunked disenchantment thesis. However, as the history of the hanged man's hand and blood-drinking demonstrates, the policies of the modern state could inadvertently promote and enable popular magical and medical practices, as well as hamper and suppress them. The enthusiasm for blood-drinking was promoted in the nineteenth century by the shift to execution practices that were thought to be more humane. The hanged man's hand continued into the mid-nineteenth century due, in part, to pragmatic administrative decisions about ensuring a limited pool of hangmen were kept happy. Some traditions declined as an inadvertent consequence of changing execution practices, rather than popular rejection or authoritarian suppression.

NOTES

1. Cited in Richard D.E. Burton, *Blood in the City: Violence and Revelation in Paris, 1789–1945* (2001), p. 101.
2. Kathy Stuart, *Defiled Trades and Social Outcasts: Honor and Ritual Pollution in Early Modern Germany* (Cambridge, 1999), p. 225; Spierenburg, *Spectacle of Suffering*, pp. 188–190. See also, Lisa Silverman, *Tortured Subjects: Pain, Truth, and the Body in Early Modern France* (Chicago, 2001).
3. *El Globo*, 23 December 1906.

4. Jules Drivon, 'Histoires de Bourreaux', *Revue d'Histoire de Lyon* 11 (1912), p. 194.

5. Carøe, *Bøddel og Kirurg*, pp. 43–44, 49–50.

6. Michael Ramsey, *Professional and Popular Medicine in France 1770–1830: The Social World of Medical Practice* (Cambridge, 1988), pp. 87, 89.

7. Frédéric Armand, *Les bourreaux de France: du Moyen âge à l'abolition de la peine de mort* (Paris, 2012), p. 214.

8. H. Sanson, *Sept générations d'exécuteurs 1688–1847: Mémoires des Sanson* (Paris, 1863), vol. 6, pp. 158–159.

9. See, for example, F. Suárez de Ribera, *Remedios deplorados probados en la piedra lydio de la experiencia* (Madrid, 1732), p. 118.

10. David Gentilcore, *Medical Charlatanism in Early Modern Italy* (Oxford, 2006), p. 221; Piero Gambaccini, *Mountebanks and Medicasters: A History of Italian Charlatans from the Middle Ages to the Present* (Jefferson, 2004), p. 24.

11. Kathy Stuart, *Defiled Trades and Social Outcasts: Honor and Ritual Pollution in Early Modern Germany* (Cambridge, 1999), p. 157.

12. Antonio Bortolotti, *Cagliostro e l'arte di sanare nel '700* (Milan, 1995), p. 73.

13. Stuart, *Defiled Trades*, p. 158.

14. Domenico Cotugno, *A Treatise on the Nervous Sciatica* (London, 1775), pp. 107, 128. See also Sugg, *Mummies*, pp. 232–234.

15. Javier Puerto, *Medicamentos legendarios: Mito y ciencia en laterapéutica clásica* (n.d.), p. 154; Evariste Thévenin, *Entretiens populaires* (Paris, 1863), p. 114.

16. Charles Estienne and Jean Liébault, *L'Agriculture et maison rustique* (Paris, 1665), p. 181.

17. A.J.B. Parent-Duchatelet, *Hygiène Publique* (Paris, 1836), 2 vols, vol. 1, pp. 22–25.

18. Charles Londe, *Nouveaux éléments d'hygiène*, 2nd edition (Paris, 1838), Vol. 2, p. 154.

19. Drivon, 'Histoires de Bourreaux', 194; L.-H. De Martin, *Des corps gras naturels et artificiels* (Paris, 1869), p. 187; *Journal de chimie médicale*, 4th S, Vol. 7 (1861), 100.

20. Franco Mormando, *The Preacher's Demons: Bernardino of Siena and the Social Underworld of Early Renaissance Italy* (Chicago, 1999), p. 96; Jean Baptiste Thiers, *Traite des superstitions* (Paris, 1697), Vol. 1, p. 380.

21. Hermann Leberecht Strack, *The Jew and Human Sacrifice: Human Blood and Jewish Ritual. An Historical and Sociological Inquiry* (New York, 1909), p. 73; Gary K. Waite, *Heresy, Magic and Witchcraft in Early Modern Europe* (Basingstoke, 2003), p. 135.

22. María Tausiet, *Urban Magic in Early Modern Spain: Abracadabra Omnipotens* (Basingstoke, 2014), pp. 93–94.
23. Evans, *Rituals of Retribution*, pp. 93–94; Keller, *Scharfrichter*, p. 232.
24. Strack, *The Jew and Human Sacrifice*, p. 74.
25. Sarah Tarlow, 'The Technology of the Gibbet', *International Journal of Historical Archaeology* 18 (2014), 681, 696.
26. Elizabeth Hurren, *Dissecting the Criminal Corpse: Staging Post-Execution Punishment in Early Modern England* (Basingstoke, 2016), p. 258.
27. Johann Jacobi Waldschmidt, *Operum medico-practicorum* (n.p., 1717), Vol. 1, p. 521; 'Physician, Apothecary, and Hangman', *British Medical Journal*, 27 December 1913, p. 1641.
28. *Doncaster Gazette*, 24 June 1870; *Leeds Mercury*, 2 February 1904.
29. *Edinburgh Evening News*, 2 January 1886.
30. Deborah Lutz, *Relics of Death in Victorian Literature and Culture* (Cambridge, 2015), pp. 97–100.
31. *The East Anglian*, N.S. 1 (1885–1886), p. 295.
32. *Medical Times and Gazette*, 1 November 1855, pp. 443–444.
33. *Leeds Mercury*, 3 February 1904; J. Turner Horsfall, *Yorkshire Notes and Queries* (Bingley, 1888), Vol. 1, pp. 63–64.
34. Davies, *Murder, Magic, Madness*, p. 34.
35. H.B. Irving, *Studies of French Criminals of the Nineteenth Century* (London, 1901), pp. 141–151.
36. *Le Figaro*, 24 November 1884.
37. *Les mémoires de M. Goron: Ancien chef de la Sûreté* (Paris, 1897), pp. 179, 178. See also, G. Variot, 'Remarques sur l'autopsie et la conformation organique du supplicié Pranzini et sur le tannage de la peau humaine', *Bulletins et Mémoires de la Société d'anthropologie de Paris* 10 (1929), pp. 42–46.
38. Paul Eudel, *L'hotel Drouot et la curiosité* (Paris, 1887), p. 249.
39. *Le Figaro*, 17 November 1886.
40. http://web.prm.ox.ac.uk/amulets/index.php/saints-amulet1/.
41. *The Works of Victor Hugo: Things Seen and Essays*, vol. 14 (New York, 1907), p. 241.
42. *Royal Leamington Spa Courier*, 19 April 1845.
43. Francisque Michel, *Le Pays Basque: Sa population, sa langue, ses moeurs* (Paris, 1857), p. 160.
44. *New-England Galaxy*, 18 June 1819.
45. John Ashton, *Social England under the Regency* (London, 1890) vol. 2, p. 60. Our thanks to Alice Dolan and Anne Murphy for this reference.
46. *Morning Post*, 28 November 1821. Our thanks to Naomi Clifford for this reference. On Norris, see Clifford's forthcoming book *Unfortunate Wretches*.

47. Owen Davies and Francesca Matteoni, 'A virtue beyond all medicine': The hanged man's hand, gallows tradition and healing in eighteenth- and nineteenth-century England', *Social History of Medicine* 28 (2015), 686–705: http://shm.oxfordjournals.org/content/early/2015/05/01/shm.hkv044.full.pdf+html.

48. See Francesca Matteoni, 'The Criminal Corpse in Pieces', *Mortality* 21 (2016), 1–12.

49. Oppelt, *Über die Unehrlichkeit des Scharfrichters*, pp. 753–754.

50. Hornum, 'The Executioner', pp. 70–71.

51. *Sandusky Register*, 10 March 1851.

52. J. Ross Browne, *An American Family in Germany* (New York, 1867), p. 132. The case is also noted in Eduard Stemplinger, *Antike und Moderne Volksmedizin* (Dieterich, 1925), p. 61; Albert Hellwig, *Verbrechen und Aberglaube; Skizzen aus der volkskundlichen Kriminalistik* (Leipzig, 1908), p. 67.

53. Strack, *The Jew*, p. 72; *New Albany Daily Register*, 10 November 1857.

54. Nils Helger, *Hällefors socken: hembygdsbeskrivning i historisk framställning* (Falund, 1945), p. 74.

55. Evans, *Rituals*, pp. 90, 93. Albrecht Keller, *Der Scharfrichter in der Deutschen Kulturgeschichte* (Hildesheim, 1968), pp. 232, 217.

56. Llewellyn Lloyd, *Peasant Life in Sweden* (London, 1870), p. 159.

57. *Northampton Mercury*, 30 August 1856. Archival material on the case can be found at http://boyebanden.dk/mordet-p-niels-hansen/eksekvering-af-dommen-1856.html.

58. *Aftonbladet*, 12 August 1846.

59. Shaun R. McCann, A History of Haematology: From Herodotus to HIV (Oxford, 2016), p. 4.

60. D. Cassano and C. Colucci d'Amato, '"The moon" and "the blood": Two emblematic symbols in headache and epilepsy according to scientific traditions of the Salerno Medical School and popular medicine in southern Italy', *Journal of the History of the Neurosciences: Basic and Clinical Perspectives* 1 (1992), 97–110.

61. *A Treasure of Health. By Castor Durante Da Gualdo … Translated out of Italian* (London, 1686), p. 46.

62. Edelgard E. DuBruck, 'Theophrastus Bombastus von Hohenheim, Called Paracelsus: Highways and Byways of a Wandering Physician (1493–1541)', *Fifteenth-Century Studies* 25 (2000), p. 5.

63. Keller, *Der Scharfrichter*, p. 232; Evans, *Rituals*, pp. 90–91.

64. Hermann Leberecht Strack, *The Jew and Human Sacrifice: Human Blood and Jewish Ritual. An Historical and Sociological Inquiry* (New York, 1909), pp. 70–71; U. Brodersen, 'Henrettet i Aabenraa'; http://den-gang.dk/artikler/2171.

65. L. Lloyd, *Peasant Life in Sweden* (London, 1870), p. 159.

66. Wolfgang Oppelt, *Über die Unehrlichkeit des Scharfrichters* (Lengfeld, 1976), pp. 752–53; Evans, *Rituals*, p. 91.

67. Nordiska Museet Arkivet, 25840. See also, for executed criminals' blood as a cure for epilepsy: NMA 30802, 30803, 23500, 28800, 23504.

68. J. Schröder, *Zoologia*. Trans. Anon (London, 1659), p. 48. See also, Sugg, *Mummies*, pp. 56–57.

69. Francesco Paolo De Ceglia, 'The Blood, the Worm, the Moon, the Witch: Epilepsy in Georg Ernst Stahl's Pathological Architecture', *Perspectives on Sciences* 12 (2004), 1–28.

70. William Cullen, *First Lines of the Practice of Physic* (Edinburgh, 1791), vol. 3, p. 365; Davies and Matteoni, '"A Virtue beyond all medicine"', 7.

71. Evans, *Rituals*, p. 92.

72. Browne, *American Family*, pp. 132–133.

73. Arthur Mitchell, 'On Various Superstitions in the North West Highlands and Islands of Scotland, especially in Relation to Lunacy', *Proceedings of the Society of Antiquaries of Scotland* 4 (1863), p. 273.

74. Ernst Ludwig Rochholz, 'Gold, Milch und Blut: Mythologisch', *Germania: Vierteljahrsschrift für deutsche Altertumskunde* 7 (1862), 414; Albert Hellwig, *Verbrechen und Aberglaube*, pp. 67–68. See also, Wolfgang Oppelt, *Über die Unehrlichkeit des Scharfrichters* (Lengfeld, 1976), p. 753; Oskar von Hovorka and Adolf Kronfeld, *Vergleichende Volksmedizin; eine Darstellung volksmedizinischer Sitten und Gebräuche, Anschauungen und Heilfaktoren, des Aberglaubens und der Zaubermedizin* (Stuttgart, 1908), p. 217.

75. See the website: http://www.genealogi.se/avrattade-start.

76. Ingjald Reichborn-Kjennerud, *Vår gamle trolldomsmedisin* (Oslo, 1947), vol. 5, p. 65; Ernst Westerlund Selmer, *Festskrift til Hjalmar Falk: 30. desember 1927, fra elever, venner og kolleger* (Aschehoug, 1927), p. 30.

77. Strack, *The Jew and Human Sacrifice*, p. 73.

78. *Catholic Telegraph*, 25 September 1852; Evans, *Rituals*, p. 92.

79. Evans, *Rituals*, pp. 91–92; Oppelt, *Über die Unehrlichkeit des Scharfrichters*, p. 754.

80. N. Alpaghi, 'Il sangue dei giustiziati in Spagna', *Archivio delle tradizioni popolari* 13 (1894), 453.

The Places and Tools of Execution

Abstract The materiality of gallows' sites and the apparatus of execution had their own magical and medical afterlife, inextricably linked with, yet ultimately independent of, the executed criminal. This chapter includes discussion on European gallows traditions, English legends of providential strangulation, and the trade in and lore of the gallows mandrake. It then focuses on the trade in hanging ropes in Europe and America, and the relationship between mementos and magical talismans.

Keywords Gallows · Providence · Mandrake · Hanging rope · Talisman

We move now from matters concerning the identity and potency of criminal bodies to their post-mortem relationship with the immediate physical environment where their last sentient moments were extinguished and witnessed. The materiality of gallows sites and the apparatus of execution have their own magical and medical afterlife, inextricably linked with, yet ultimately independent of, the executed criminal.

During the eighteenth and nineteenth centuries, execution sites across Western Europe were increasingly situated in fixed locations, with those sentenced for capital offences being condemned to die 'at the usual place'. There had always been a mix of urban and rural gallows, of course, but the process of state-building led to centralised control over

© The Author(s) 2017
O. Davies and F. Matteoni, *Executing Magic in the Modern Era*,
Palgrave Historical Studies in the Criminal Corpse and its Afterlife,
DOI 10.1007/978-3-319-59519-1_4

the act of execution as punishment and spectacle. It also led to the closer spatial association of gallows with judicial institutions, such as jails, rather than with the communities in which the crimes took place or where the condemned criminals lived.[1] There are, of course, exceptions and anomalies. In England, between 1720 and 1830, nearly 200 people were hanged on specially erected scaffolds near the scene of their crime, quite often on remote hilltops.[2] Their purpose was to act as exceptional spectacular warnings—particularly to rioters—and to spread the message of law and order in often isolated rural areas.

There was also a general shift away from permanent gallows in the modern era. The famous Tyburn gallows in London was replaced in 1759 by a portable scaffold. In Lower Austria, a law of 1786 ordered the removal of a series of permanent, monumental stone gallows that had been constructed in the seventeenth and early eighteenth centuries, usually on natural elevations near major roads.[3] In 1808, a German Police chief complained to his superiors in Munich that these 'misshapen monuments, gallows, and scaffolds still insult the bright eye of the wanderer and carry his mind involuntarily back to the days when the hangman was one of the state's busiest servants'. He continued that 'it is incompatible with the principles of the present to have the gallows as the first thing one sees on approaching every important place'. Three years later, the kingdom of Württemberg dismantled its permanent gallows and ravenstones, and other German states soon followed suit if they had not done so already.[4] The disused gallows that remained in the landscape shifted from being landmarks of state justice to being antiquarian curiosities. A visitor to the Swedish town of Visby in the early nineteenth century remarked, for instance, upon the 'relic' of the town's disused permanent gallows, 'yet remaining, formed of the pillars placed in a triangular position; a beam serving as the gallows passes from pillar to pillar; below is a well, into which the bodies of the unhappy sufferers were either thrown or suffered to fall piecemeal'.[5]

The terms gibbet and gallows are often confused, or used indiscriminately for both execution and post-mortem sites. In some cases, such as with English crime scene executions, the same post was used for both execution and hanging in chains, thereby leaving a much longer physical reminder of both judicial acts. In rural Southern Netherlands, execution gallows, post-mortem display and eventual burial were usually on the same spot, as distinct from urban areas where executions took place in the town centre, with the corpse then being taken outside the town for display. As Sarah Tarlow explores, gibbets remained striking features in the landscapes for decades: 'The presence of a gibbet could change the experience of a local landscape for a long time after its erection even to the present day. ... giving emotional impact to local journeys.' In the eighteenth century, gibbets were still sometimes located at places that might already have their own folklore, such as on old prehistoric barrows, which were usually sited on hilltops and ridges, or crossroads or parish boundaries where medieval and early modern executions had long taken place.[6]

So, a few decades before public execution was abolished across much of Europe, the sight and site of the gallows had become largely fleeting urban fixtures. Yet across the countryside, the legacy of centuries of executions remained in place names, such as Hangman's Field and Gallows Hill—*Galgenberg* in German. Sometimes place names provide the only evidence of the location of temporary or one-off execution and gibbeting sites over the centuries. Some can be dated in the archives as having medieval and early modern origin, but other names record more recent events, such as Gibbet Hill Lane at Scrooby, which memorialises the 1779 gibbeting of John Spencer.[7] More to the point, we find numerous stories, beliefs and legends relating to the ghosts of the executed haunting the locations long after the gallows or the gibbet had been taken down. Nearly two centuries on from the gibbeting of a man on Maundown Hill, Wiveliscombe, Somerset, in 1727, for murdering his mother, Thomazin Tudball, people still told stories of his frightening spiritual presence as they drew near the isolated spot. Horses became terrified as they approached where the gibbet used to stand.[8] Accounts of horses sensing and fearing old gallows and gibbet spots were quite widespread. It was recorded in the early twentieth century, for instance, that the reason horses feared crossing Pilfer Bridge, near Normanby, Lincolnshire, was because of an old long-gone gibbet nearby, memorialised in the locality by a property called Gibbet Post House.[9]

Writing in the 1890s, the observant Cumbrian antiquarian, Henry Swainson Cowper, mentioned the folklore surrounding Gibbet Moss, near the Lake District town of Hawkshead. The poet, Wordsworth, recalled seeing the gibbet post still standing there as a boy. As recorded in the parish register, it was that which bore the body of Thomas Lancaster, who was executed in 1672 for poisoning his family. He was hanged in front of his home and then his corpse was taken away to be hanged in chains on the spot that came to be known later as Gibbet

Moss.[10] As Cowper noted, 'The simplest form of haunt is simply that in which a particular place bears a bad reputation without there being attached to it any distinct apparition.' Other than its bad reputation and name, Cowper thought 'probably there are now many people who are unaware that a gibbet ever stood here.' This seems to be the case for most haunted gibbets recorded in the late nineteenth and early twentieth centuries. The post long gone, the name of the executed usually forgotten in local lore, no one could visualise what he might look like: but there remained an unsettling presence.

The gallows and gibbet from which swung the corpse of Thomas Colley, executed for the drowning of a suspected witch in 1751, were haunted by a black dog. Colley was tried in Hertford, but he was executed and gibbeted thirty miles away at Gubblecote Cross, which was a crossroads near where the fatal swimming occurred at Marston Mere. The execution and gibbeting did not take place at the crime scene, because the locals successfully petitioned against it.[11] Just over a century later, the local schoolmaster claimed to have seen the demonic dog when returning home one night in his gig. 'When we came near the spot, where a portion of the gibbet had lately stood, we saw on the bank of the roadside ... a flame of fire as large as a man's hat ... I then saw an immense black dog lying on the road just in front of our horse ... The dog was the strangest looking creature I ever beheld.'[12] The association of animal spirits with humans who died a violent or tragic death, such as the executed and suicides, was widespread in folk tradition.[13] We will return to this in the next chapter.

PROVIDENTIAL GALLOWS

Dotted across the English landscape are some two dozen or so 'hangman's stones' that also served as symbolic warnings to would-be thieves. These are usually Bronze Age standing stones, wayside crosses, or boundary and milestones. Writing in 1803, the Gloucestershire antiquarian, Thomas Rudge, wrote of one such example in the county, 'An ancient rude stone, about four feet high, commonly called Hangman's stone, with a vulgar tradition annexed to it, not worth recording, stands in the road, about two miles from Cirencester, and is now converted into a mile-stone.' The vulgar tradition referred to was related by an earlier, and less pompous, Gloucester antiquarian, who explained that it was called Hangman's Stone because 'it is said, a fellow resting a sheep thereon, (which he had stolen, and tied its legs together for the convenience of carrying it) was there strangled, by the animal's getting its legs round his neck in struggling'.[14] The same story of the asphyxiation of a sheep thief resting upon a stone with his ill-gotten carcass was attributed to most of the other hangman's stones, though a mid-nineteenth-century poem records the similar fate of a Leicestershire deer poacher named John of Oxley:

There was Oxley on one side the stone,

On the other the down-hanging deer;

The burden had slipp'd, and his neck it had nipped;

He was hanged by his prize – all was clear!

The gallows still stands upon Shepeshed high lands,

As a mark for the Poacher to own,

How the wicked will get within their own net;

And 'tis still cal'd 'THE GREY HANGMAN'S STONE.'[15]

The fact that the same legend was attributed to different stones around the country was remarked upon in the late nineteenth century, and was given further attention in the 1920s by the influential archaeologist, O.G.S. Crawford, who compiled a brief gazetteer of sites.[16] Crawford mused on the possible medieval origin of the hangman's stone, being so named from being parish or hundred boundary stones near a gibbet or execution site. The strangled sheep-thief story occurred much later to explain the name at a time when the execution site had long fallen out of public memory. Leslie Grinsell expanded on Crawford's list of sites, and, as far as it is possible to date, charted the chronology of the tradition. He found a couple of seventeenth-century references, but the rest dated from the mid-eighteenth century, the bulk being from the 1830s onwards. This dating only concerns the first literary references to the tradition of the various sites, of course, and does not necessarily reflect the age of the tradition in oral circulation.

How this 'migratory legend', as folklorists call it, became so widespread is difficult to pin down. The legend clearly relates to the fable, '*Poena sequens*' or 'Punishment following', which appeared in

sixteenth-century emblem books. A woodcut of a thief, being strangled by the tied legs of the pig carcass he has just stolen as he lies down
to rest against a balustrade, is first depicted in Johannes Sambucus's
Emblemata (Antwerp 1564). This was a book of images and associated moralistic epigrams that Sambucus intended 'to instruct and to
delight'.[17] It was soon translated into French, Dutch and English,
and reprinted in other books of emblems. It first appears in English in
Geoffrey Whitney's *A Choice of Emblemes* (1586), though in his version
the thief is carrying a sack of meat stolen from a shambles.

> The heavie loade, did weye so harde behind,
>
> That whiles he slept, the weighte did stoppe his winde.
>
> Which truelie shows, to them that doe offende,
>
> Although a while, they scape theire just desertes,
>
> Yet punishment, dothe as theire backs attende ...
>
> And thoughe longe time, they doe escape the pikes,
>
> Yet soone, or late, the Lorde in justice strikes.[18]

Did Sambucus take his story from a widespread European legend circulating in oral form? Or was it his literary invention that then, rather like
some fairy tales, found its way from the page and into English oral folk
tradition? A search through English ballads that might have borrowed
Whitney's verse, and thereby disseminated the legend, did not turn up
any examples.

Grinsell suggested that the apparent predominance of this English
folk tradition from the mid-eighteenth century onwards was a reflec-
tion of the heightened concern over sheep stealing at the time. This
was expressed in 1741 by *An Act to render the Laws more effective for
the preventing the stealing and destroying of Sheep*. Despite the periodic
concern, particularly when dearth and poor harvests led to apparent
spikes in sheep theft, sheep stealing was considered a less heinous crime
than horse theft and the punishment was often mitigated, particularly
to transportation. Still, the deterrent effect was considered effective. In
1790, one judge expressed his satisfaction that since he 'ordered the execution of two sheep stealers 2 or 3 years ago the offence ... which had
filled the calendar of every circuit for several years before that example
was made has not once appeared'.[19]

The divine nature of the sheep thief's demise was suggested in one of the earliest versions of the legend associated with the Hanging Stone near Combe Martin, Devon. Referring to it, Thomas Fuller opined in the 1660s that 'it appeareth rather a *Providence*, than a *Casualty*, in the just execution of a *Malefactor*.[20] The providential aspect of John of Oxley's demise is certainly inferred in the poem above.

The death penalty for sheep stealing was repealed in 1832, and it has been suggested the nineteenth-century ubiquity of this morality legend may have been a popular response to the ending of capital punishment for the offence,[21] or it may just be a question of source bias. The hangman's stone tradition is best understood in the context of the broader and older tradition of divine justice or retribution being memorialised in the landscape, which was prevalent in folklore and ballads from the Reformation to the twentieth century.[22] Another example is the legend associated with various Bronze Age stone circles and standing stones, that they were sinners petrified for dancing or playing games on the Sabbath. We see, in these traditions, popular manifestations of the state's attempt to imprint morality, justice and punishment on the landscape.

THE GALLOWS MANDRAKE

In early modern and modern Europe, there was an oft-recorded belief that the mandrake, a plant with well-known magical associations, grew under the gallows. Ancient Greek writers, such as Hippocrates, noted the beneficial and toxic properties of the plant. The main source for its enduring magical associations, though, was an account of the plant written in the first century CE by the Roman-Jewish historian, Josephus. He told how the plant would not 'yield itself to be taken quietly, until either the urine of a woman, or blood, be poured on it'. Josephus was also the source of the method of pulling the mandrake out of the ground by tying a dog to the plant, the dog dying in the process. The association of the mandrake with the gallows is much later, though. It is first recorded in 1532 by the physician and botanist, Otto Brunfels (1488–1534), although he refers to the eleventh-century Persian physician, Avicenna, as a source, though no one has yet found it mentioned in Avicenna's work. What Brunfels stated as a current belief was that the mandrake, or *alraun* in German, grew from 'the sperm of a urinating thief' on the gallows.[23] Over the next few years, the notion appeared in a few other German books. In 1534, Leonhard Fuchs referred to the 'land swindlers' who carved roots into human form, 'and they lie even more by saying that these roots have to be gathered under the gallows, with many ceremonies'.[24] It is probably due to the influence of Brunfels work, though, that by the seventeenth century this notion, and variations upon it, were widespread in European medical literature. The French apothecary and herbalist, Laurent Catelan (c. 1568–1647), for instance, wrote a study of the mandrake, in which he discussed the generation of the mandrake from sperm released from criminals hanged or crushed on the wheel. Because of the frequency of hangings, he explained, the semen fell on ground that was rich in human fat, 'like that in a graveyard', which aided the fertilisation process. A woman's sperm (it was believed by Galenists that women also produced semen, though less potent) would not generate mandrake, however, 'even if she was strung up or crushed'.[25] Small variations appear in the various seventeenth-century sources, such as that the mandrake was only generated from the fluids of a man unjustly hanged for theft, or who was a virgin, or a hereditary thief.[26]

While the medicinal properties of mandrake root were known widely enough in British herbal medicine, where it was used in particular to promote conception, the gallows mandrake tradition circulated little in English folk belief.[27] John Gerard (1545–1612), in his influential *Herball*, castigated the 'ridiculous tales' regarding the plant, ruminating whether they were broadcast by 'old wives, or some runnagate Surgeons or physicke-mongers'. He was categorical that it was 'never or very seldome' found to grow naturally under the gallows.[28] Likewise, Thomas Browne's dismissal of the notion in 1685 as erroneous and 'injurious unto Philosophy' was a response to classical and continental sources, rather than native tradition.[29]

In Iceland, the mandrake was known as 'thieves root' and was thought to grow from the froth that dribbled from the mouth of the hanged. However, the gallows mandrake tradition was strongest in German lands, where it was known as the *galgenmännlein* or 'little gallows man', generated from the urine or semen leaking from the male corpse. The plant was also tellingly associated with crossroads where suicides were buried. While much of the evidence for the early modern tradition was regurgitated from a narrow pool of medical literary sources, it is clear that there was a popular trade in mandrakes purportedly taken from gibbet sites. The seventeenth-century German poet, Johann Rist, for instance, wrote of seeing an old carved mandrake placed in a little coffin with a cloth cover depicting a thief on the gallows.[30] Paracelsus complained of the cheating vagabonds who traded in mandrakes, stating that mandrake roots did not grow in the form of humans; it was unscrupulous people who carved them thus. Brunfels also grumbled about those who cheated people by carving bryony roots and passing them off as mandrakes. In the late sixteenth century, Frantz Schmidt, the German executioner-physician and diarist, noted the activities of the thief and card sharp, Hans Meller, who traded in magical objects. He was convicted for coating yellow turnips in fat, sticking hair on the top and selling them as mandrakes.[31]

That some executioners were involved in the mandrake trade—selling fake products or otherwise—is evident from a letter written in 1575 by a burger of Leipsic to his brother in Riga. His brother was experiencing a series of misfortunes: his livestock was dying mysteriously, his beer had turned sour, and so had his relations with his wife.

I have, therefore, gone to those who understand such things to find what is needed, and have asked them why thou art so unlucky. They have told me that these evils proceed not from God, but from wicked people; and they know what will help thee. If though hast a mandrake and bring it into thy house, thou shalt have good fortune. So I have taken the pains, for thy sake, to go to those who have such things, and to our executioner have paid sixty-four thalers and a piece of gold drinkgelt to his servant, and this [mandrake], dear brother, I send thee.

Instructions were included about how the mandrake had to be left to rest for 3 days before being soaked in warm water, which was then to be sprinkled on the sills of the house and the sick animals.[32]

In eighteenth- and nineteenth-century France, interest in the mandrake was fuelled by the popular book of magic, *Secrets merveilleux de la magie naturelle et cabalistique du Petit Albert*. This included several pages on how to grow the *mandragore* or mandrake, and noted the belief that it brought good fortune to the possessor. The 'author' recalled meeting a rags-to-riches peasant, who told him how his own wealth was due to a gypsy woman who had told him how to grow a mandrake. This involved taking a bryony root, which had a human form, from the earth during the spring, when the moon was in an auspicious conjunction with Jupiter or Venus. The root then had to be trimmed and planted in a cemetery or in the grave of a dead man. Every day for a month, at sunrise, it had to be watered with the whey from cows' milk in which three bats had been drowned. After 1 month, the root was dug up and would be found to have taken on a human likeness. It then had to be dried in an oven with the plant vervain, and finally wrapped in a piece of the burial shroud of a dead person.[33] So here we have an essential association with places of death, if not explicitly with the gallows. The *Petit Albert* also includes the most culturally influential recipe for creating the notorious Hand of Glory, in French known as the *Main de Gloire*, which is thought to be a corruption of the French for mandrake—*mandragore*. How the etymology arose is confused, but there is an obvious shared association with the gallows corpse.

Come the nineteenth century, it would seem that the gallows mandrake had become largely a matter of fiction rather than an object of the magic trade. The legend appealed to the imaginations of the Romantics.[34] Ludwig Achim von Arnim's German novella, *Isabella of Egypt, the First Young Love of Emperor Charles V* (1812) made particular play with the traditions. The father of the eponymous Isabella, who was the king of the gypsies, is unjustly hanged as a thief. She reads his books of magic and comes across a ritual for raising a homunculus from a mandrake. She goes to the gibbet site where her father was hanged to gather one, sacrificing her pet dog to pull it out of the ground. In Britain, Harrison Ainsworth's popular novel, *Rookwood* (1834), which was set in 1737, also figures gypsies as well as the highwayman, Dick Turpin. In one passage, one of Turpin's accomplices sings a ballad about the mandrake that begins with 'The mandrake grows 'neith the gallows tree', with a subsequent verse explaining:

> At the foot of the gibbet the mandrake springs,
>
> Just where the creaking carcass swings;
>
> Some have though it engendered
>
> From the fat that drops from the bones of the dead;
>
> Some have thought it a human thing;
>
> But this is vain imagining.

Jacob Grimm's account of the *alraun* in his magnum opus, *Teutonic Mythology* (1835), further promoted the gallows mandrake from legend to myth, by suggesting that the plant was associated with the goddess, *Aurinia*, and that early Germanic wise women/diviners, known as the *alrúna*, gave their name to the mandrake. It was sources such as this that inspired the unsettling vampire novel, *Alraune* (1911), by Heinz Ewes. It concerns an anthropologist named Frank Braun, who, on hearing of the legend of the gallows mandrake, convinces his maverick scientist uncle to recreate a homunculus by artificially inseminating a Berlin prostitute with the semen of a notorious hanged murderer. The result is a beautiful but deadly female vampire called *Alraune*. The story was made into several film adaptations over the next few decades.[35]

It is difficult to demonstrate any firm correlation between the decline of the gallows mandrake trade and changing execution practices and patterns, but there were some likely influences. The shift of execution sites to centralised urban locations and the decline of post-mortem display in the countryside would have, practically and notionally, reduced the association with the earthy human compost in which the mandrake was thought to grow. Mandrakes without gallows associations continued to be sold, however, as a herbal remedy and as an amulet for good luck, into the nineteenth and early twentieth centuries. The London folklorist, Edward Lovett, purchased a couple of specimens that had not been 'touched up', and recorded that people sometimes fixed one to the bedhead for good luck. In the early twentieth century, a vendor of what were purported to be mandrakes with humanoid forms sold them from his pitch in Petticoat Lane, London. He told his customers, 'I pulls 'em at midnight, and they screams 'orrible as they come out of the ground.' Penny slices of mandrake were also sold as a general cure-all.[36]

THE LAST OF THE EXECUTIONERS' PERQUISITES

As has already been noted, the scaffold or execution platform was a temporary construction. However, while it still stood, there was a brief opportunity to obtain part of the structure. By the mid-nineteenth century, the trade in fresh body parts was at an end across Europe, the control over the corpse having been wrested from the executioners, but in hanging countries, in particular, they still had a degree of control over the gallows and the tools of the trade, and people had long sought out pieces of the execution apparatus for medicine and magic.

Nails were widely prized in some places, for example. In 1776, a Swedish executioner, named Lars Hierpe, was prosecuted for lending the nails that had been driven into the head and hand of an executed thief to a farmer, named Jöns Persson, who used them to cure his livestock.

Hierpe testified that he had taken the nails from the body, 'partly because of one Jöns Persson's wish, partly because of the old tale that such nails are useful for the horse-barterer to make the horses livelier'.[37] Thomas Browne, writing in the mid-seventeenth century, mentioned how 'for amulets against Agues we use the chips of Gallows and places of execution', and the late-eighteenth century antiquarian, Francis Grose, noted that pieces from the gibbet or gallows were placed in a bag and worn around the neck to cure or prevent ague. Earth from the scaffold graves of the executed was prized for its miraculous power in parts of Italy. The antiquarian, John Brand, recalled having seen, in 1746, blood-soaked scaffold sawdust from the beheadings of the treasonous Jacobites, Lord Balmerino and the Earl of Kilmarnock, used to charm away some illness. Jacob Grimm referred to the belief that a spur made from a gibbet chain possessed magical potency.[38]

The most enduringly desired artefact from the execution site, and the one most controlled by the executioner, was the hanging rope. After all, it was, by custom, the hangman who paid for it from his fees. It had been much sought after for centuries for healing. Before the age of the guillotine and garrotte, the hanging rope was used in France and Spain as a cure for headaches by touching the temple with it, and pieces were kept in the pocket against toothache. In Transylvania, a piece of hangman's rope placed in the bed was thought to prolong the life of those near death.[39] The rope, like criminal corpse body parts, was also thought to have non-medical magical properties. The inquisition tribunal records for sixteenth-century Spain contain numerous references to its use in love magic. One woman confessed before the tribunal 'that she used rope from a hanging and carried it with her to lure men's wills'. María Tausiet has observed 'the symbology of subjugation linked to death by asphyxiation and embodied in the hangman's noose, which had the ultimate control over another person's will'.[40] In early modern Germany and in early nineteenth-century Molise, central Italy, it was thought to act as a talisman against gunshot. The most pervasive and enduring market for hanging rope through to the twentieth century was amongst gamblers of all levels of society, who prized it as a luck-bringer.[41] As with body parts, the executioner had to contend with competition for this perquisite. In one early modern German instance, a robber, named Hörnlein, was prosecuted for having stolen a piece of rope from the gallows in broad daylight.[42] There was no doubt a trade in fake pieces of rope as well.

The rope was a prominent part of British folk medical and magical tradition, as it was the main hanging country in Western Europe during the nineteenth century. Just as people came to hangings to feel the touch of the hanged man's hand for swellings, so they also came to obtain a bit of the rope for the cure of ague, epilepsy and other ailments. One woman travelled fifteen miles to attend the hanging of the highwayman, Nicholas Mooney, in 1752, and so obtain a section of the rope. John Brand remembered attending a hanging at Newcastle later in the century and witnessing several men climb the gallows after the body had been cut down to obtain pieces of the cord that remained.[43] On most occasions, access to the rope was carefully controlled by the hangman. A nineteenth-century Herefordshire woman, whose child suffered from scrofula, was advised to go to the next hanging at Hereford and request the hangman to pass the rope over his body and lash him with it.[44] The most common practice was for the hangman to cut up the rope shortly after the execution and sell the pieces then and there. The rope that hanged Joseph Wall, who was executed in London for flogging a soldier to death in 1802, was sold for one shilling per inch. There was a lively trade at the execution of the murderer, Benjamin Ellison, at Bodmin gaol, Cornwall, in August 1845. The hangman, George Mitchell, charged a shilling a piece. Some months later, a surgeon at the local infirmary was treating a patient for an ulcerous wound on his back and found a little chintz bag around his neck. When he inquired what it contained, the patient replied that it was a bit of the rope from the recent hanging at Bodmin. The last time he had had a problem with ulcers on his back, the patient had buried a bit of the rope, and as the rope had rotted so his wound had healed. Now he kept a piece on him permanently.[45] Any rope that did not sell on execution day could be traded subsequently, of course. A month after the notorious poisoner, William Palmer, was hanged at Stafford prison in June 1856, a Scottish newspaper reported that a man from Dudley, where the hangman George 'Throttler' Smith resided, had recently arrived in Lochmaben, Dumfriesshire, selling pieces of the rope at five shillings an inch.[46]

Just as the authorities made a concerted effort to stamp out the hanged man's touch during the mid-nineteenth century, so we find attempts to suppress the executioners' hempen perquisite. The rope used to hang William Dove in York, in 1856, was ordered to be placed in the coffin. Victor Hugo related that after the execution of Tapner in 1854, 'the dead man was cut down in an hour, and then it was a question who should steal the cord. The assistants threw it down, and each one claimed a piece; but the sheriff took it and threw it in the fire.' But people still came to collect the cinders.[47] Things got decidedly awkward in British Guyana in 1871, when the colonial authorities forbade the hangman from selling the rope. He had been accustomed to charging between four pence and a shilling a piece. At the next execution in the colony, that of a wife murderer, the hangman protested at the loss of his perquisite by refusing to cut down the dangling corpse.[48]

The Vienna *Scharfrichter*, Heinrich Willenbacher, who executed a modest 36 people during his 24-year tenure during the mid-nineteenth century, reportedly pined for the old days when Austria had its headsmen and their noble skill with the sword. 'There was room for talent in those days,' he was apparently wont to lament. 'But now it is poor work. Rope, rope, nothing but rope! What's hanging, indeed?'[49] He kept the official four-foot-long execution sword above his bed to remind him of better times, but it was rope that sustained his livelihood. Willenbacher was prosecuted in 1881 for quackery and abusing his position.[50] During the prosecution of a recidivist thief, who was sentenced to 4 years' hard labour, his elderly mother gave testimony that she had tried everything to get her son to mend his criminal tendencies, including a visit to Willenbacher. He gave her what he claimed was a bit of the rope with which he had hanged Enrico Francesconi in December 1876. Francesconi, who was found guilty of the sensational robbery and murder of a postman, was the first to be hanged behind closed doors (by Willenbacher) in the Vienna jailhouse, following a law of 1873.[51] Willenbacher instructed the old woman to place the rope under her son's pillow, and if that did not cure her son of his criminality, then he would draw blood from him with his executioner's sword. After hearing this testimony, the district court ordered Willenbacher's arrest. He pleaded clemency, saying that with so few executions to conduct in Austria, he lived in poverty with his six children. He was evidently pardoned and continued in the role of hangman until his death in 1886, when it was reported that, despite his penurious position, the money he made by selling the rope to 'superstitious people', apparently at ten shillings an inch, was charitably distributed amongst the poor.[52]

Where public hanging continued into the late nineteenth and early twentieth centuries, we find the continued resort to the hanging rope. This was mainly in Eastern Europe. In 1880, the Russian state executioner, a convict named Froloff, who had murdered his own family, made good money from selling his ropes, including those he used to hang a group of Nihilists who had assassinated Czar Alexander II. He bungled the job badly, with the rope around Timofey Mikhaylov's neck snapping twice before Froloff finally put the poor man out of his agony. Still, this was good news for the hangman, because it left him with more rope to sell.[53] He sold several dozen pieces on this occasion at three to five roubles. Young gamblers were the keenest purchasers, but there was clearly also a relic trade, with one report stating that strips of the hoods placed over the executed were sold to Nihilist supporters for a healthy profit.[54] Following an execution in Bulgaria in 1915, it was reported that there was a crush inside and outside the prison due to people attempting to obtain pieces of the rope for luck and good health.[55] A decade later, there were chaotic scenes after the public hanging of three Communists found guilty of blowing up St. Nedelya Church in Sofia. Petar Zadgorski, Marko Fridman and Georgi Koev were hanged separately by the country's official hangman, Hussein Jasara, an itinerant gypsy, who reaped a handsome profit selling pieces of the three ropes for good luck. For some, Jasara got his just deserts when he was killed in a fight in the gypsy quarter of Sofia in 1932.[56]

Where the sale of the rope was not expressly forbidden by the authorities, the end of public hanging did not necessarily mean an end to this hangman's privilege. When the Hungarian hangman, Michael Bali, retired in 1924 after 30 years' service, he had accumulated enough money to purchase a small estate near Budapest. Some of those funds came from his sale of pieces of the hanging rope.[57] During his time in office, he had to fight hard and publicly to try and retain his perquisite. In 1912, Bali had a widely reported dispute with the public prosecutor of Temesvár (Timişoara in Romania). Shortly after he had conducted an execution behind the prison walls, Bali untied the rope and began to cut it up, intending, as usual, to auction the pieces to crowds waiting outside. The prosecutor, who was in attendance, ordered Bali to hand over the rope to him, as riotous scenes had recently taken place in similar circumstances following a hanging at Miskolez. A Vienna correspondent, observing a hanging around the same time, probably the same, reported on the crush to get a bit of rope, and that away from the hubbub 'elegant women in carriages' waited expectantly to purchase pieces.[58] Bali refused, arguing that not only was it his perquisite by custom, but also his own property, as he had to purchase the hanging rope at his own expense. A policeman took the rope by force and Bali stated he would take legal action to recover it.[59]

MONEY FOR OLD ROPE

As already noted, in France, the Revolution had a considerable impact on the perquisites, status and medical practices of the *bourreaux*. The guillotine hit their income as they no longer had a supply of hanging rope to sell. Their Spanish brothers felt the same with the adoption of the garrotte. However, the power of the hanging rope lived on in both countries. The critic of popular errors, Jacques-Barthélemy Salgues, wrote in 1810 that the phrase '*avoir de la corde de pendu*' ('to have the rope of the hanged') had become a popular proverb to mean having constant good luck.[60] No wonder the desire for old rope continued. It was reported, in 1888, that the *corde de pendu* was still widely believed in and desired as a talisman in Bordeaux and the Gironde region, and in 1890 the folklorist, Paul Sebillot, compiled a long list of references to the ongoing potency of the rope in French, Belgian and Portuguese folklore.[61]

As well as the trade in very old pieces of French hanging rope, it could, of course, still be obtained from countries that continued to practise hanging long after France had given up the method. In 1874, the British and French press reported how a French businessman, who had visited Dublin a few years before, had taken the opportunity while there to purchase a piece of rope at the recent hanging of an Irish murderer conducted by William Calcraft. He apparently paid Calcraft five sovereigns for it, in the hope that it would bring good luck and prosperity to his family and business. The press recounted how the poor man experienced just the reverse. One of his children nearly died after 'playing hangman' with the rope. When he used the rope to seal a box of money on a business trip, he was robbed of its contents during the journey. In a state of despair, he later attempted to hang himself with the rope, only to be cut down by his wife before he died.[62] When, in 1883, the English press reported that 21 of the hanging ropes owned by the recently deceased British executioner, William Marwood, were to be sold at auction in London, one newspaper suggested that 'it is only to be feared that they have come, in the first place, to the wrong market'. There was huge demand in Paris for hanging rope as a lucky charm, it reported, noting that French gamblers and the superstitious had to rely on suicides for pieces.[63]

In the 1880s, it was apparently not unusual for Parisian medical students to sell bits of the rope that were still attached around the necks of suicides brought into the hospitals. One report tells of a piece given to a chemist's assistant in Amiens, who then went on to win one hundred thousand francs in a lottery.[64] There are examples from nineteenth-century Sweden of suicide rope also being used for curing sores and taming horses.[65] In 1923, in the village of Klimontowice, in the region of Galicia, which now straddles parts of modern Poland and Ukraine, a farm labourer, named Jan Pietraszek, was found hanging in his barn. A local bailiff confessed that he and Pietraszek had been on a drinking spree and, on returning to the latter's home, the labourer died of a heart attack. The bailiff, knowing the value of the hanging rope, hatched a plan whereby he got a cord and strung up Pietraszek's corpse to make it look like a suicide. His intention was then to obtain the rope and sell it in pieces.[66]

What of the influence of those who made hanging ropes? In early modern Brittany, the ostracism of the rope-makers was apparently unusually marked, with them being classed alongside outcast trades, such as skinners, knackers and hangmen. They turned their situation to advantage by selling charms and talismans.[67] Although only a very tiny percentage of the rope-making business was concerned with hanging rope, the association was strong, with jokes about the rope-makers' enthusiasm for this macabre aspect of their trade. There was the much-repeated story of the rope manufacturer, who complained, 'what makes it hard on rope-makers is that that least fifty men die daily of natural causes who ought to be hanged.' Much humour arose during the proceedings for bankruptcy against the English hangman, Thomas Henry Scott, at Halifax County Court in 1899. His rope-making business had failed, and government requests for his hanging services had dried up. Scott referred to himself in court as being in the 'rope trade', which was the cue for humorous exchanges as to which of his occupations he was referring.[68] As with hangmen who ended up being executed for murder, so reports about rope-makers who committed suicide by hanging themselves proved too irresistible to pass up without comment. Regarding a rope-maker who hanged himself in Stamford, for example, it was observed that it was 'a grim compliment to his peculiar calling'.[69] Some rope-makers were happy to exploit their association with death, though.

Prior to the hanging of William Dove at York, the locally commissioned rope-maker entered into a deal with a public house to display the cord, thereby attracting a large crowd of punters wishing to touch it—and hopefully buy more beer. The rope that put an end to William Palmer in the same year was commissioned from a rope-maker, named Coates, who was also a porter at Stafford station. He made thirty yards of it, with the help of other fellow station employees. The required length for the hanging was, of course, far less than thirty yards. Coates had seen a good money-making opportunity. He cut up the excess into 2- or 3-in pieces and hawked them as the rope that hanged Palmer, selling them for up to half a crown around Stafford.[70]

During the second half of the nineteenth century, one of the main suppliers of gallows ropes across America was the Edwin H. Fitler cordage works at Bridesburg, Philadelphia. Its employee, Godfrey Boger, who died in 1911, specialised in the fabrication of gallows ropes for the company, making his first one for a hanging in 1854. Only one of his ropes ever broke. Hemp of the finest quality was required, and this meant purchasing it from Italy. Boger never witnessed an execution, despite many offers, and was unconcerned by the notion that gallows rope-makers were destined to die tragically. By custom, though, the company never charged for execution ropes, only for the cost of delivery, suggesting some sensitivity to the beliefs surrounding their doleful trade. Other rope-makers abided by no such custom, with one charging five dollars for ropes with a ready-made noose.[71] In 1909, Sheriff John Schum of York County, Pennsylvania, reported having trouble finding a gallows rope in anticipation of hanging the convicted murderer, George Govogovitch. He put down his difficulty to the superstition amongst rope-makers that making ropes for hangings brought bad luck. He recalled that, some years before, when sourcing a hanging rope, several cordage companies had gone out of business, and their fate was attributed to them having branched into the gallows trade.[72]

MEMENTO OR TALISMAN? AN AMERICAN PERSPECTIVE

After scouring the folklore archives, the eminent American folklorist, Wayland Hand, was led to conclude that the use of the hanging rope as a cure was rare in America, and largely limited to Anglo-Americans and German-Americans in Pennsylvania and Ohio. Use of parts of the scaffold was equally scarce. He suggested that the lack of data on both traditions was a result of the 'rarity of hanging at a time when folklorists, local historians, and other antiquarians came on the scene to report such happenings'.[73] Public hangings had largely ended in the north-eastern states by the mid-nineteenth century, and most of the southern states by the end of the century. Arkansas continued to conduct public hangings for rapists into the early twentieth century, while Georgia and Mississippi briefly re-authorised public hanging around 1900, but as states turned to the electric chair, the noose was increasingly abandoned. Kentucky was the last to continue public hangings into the 1930s.[74] The abolition of public execution was not straightforward. Numerous hangings continued in semi-public spaces, or in jail yards, where the number of invited guests seems to have been quite considerable sometimes. Neither was solemn, respectful behaviour guaranteed, with some reports describing the selected witnesses behaving like a 'mob'.[75] So there *were* opportunities to obtain hanging rope into the early twentieth century, when folklore collecting was thriving.

As Hand suggested, searches of other sources, such as diaries, police records and 'rare local histories', might prove helpful to clarifying the practices of execution healing and magic. The Online Archive of American Folk Medicine, for example, reveals that the curative hanging rope tradition for fits and warts was also known in Maryland and Indiana.[76] The digitisation of newspaper reports helps the research into this tradition. Such reports confirm that the belief and usage of the gallows rope in popular medicine and magic was much more widespread than the folklore sources suggest. This is not really surprising, considering how widespread the demand for rope was in Europe at the same time.

The *St. Louis Republican* noted, in 1882, that 'if all the hangman's rope were taken from the pockets of the superstitious St. Louisans, they would form a rope of considerable length'. It was considered a 'cure for rheumatism, consumption, heart disease, apoplexy, and everything else' and several people in the town were selling pieces.[77] After the execution of the murderer, Jack Reynolds, in New York in 1870, an elderly lady visited The Tombs (the New York Halls of Justice and House of Detention) and asked the deputy warder for a few inches of the hanging rope to cure her son, who suffered from the King's Evil. The rope had, in fact, been kept by the carpenter who built the scaffold, so a medical man present promised to obtain a piece for her, went into the kitchen and chopped a bit of clothes' line and gave it to her, leaving the woman satisfied.[78] Just after the hanging of William Finchum for fratricide, in Harrisonburg, Virginia, in December 1887 (Virginia had been hanging before select private audiences since 1879), 'the rope was cut into small bits, and distributed to those of the spectators who wished it. Some took it from curiosity, and other to keep, on account of a superstition that it will keep off diseases, fits, spells, & c.'[79] At the New Orleans murder trial of Patrick Egan, in December 1885, it transpired that Egan, a well-known hoodlum, carried in his pocket a piece of hangman's rope for good luck. He had obtained it while attending the prison hangings of Victor Eloi and Kendrick Holland, condemned for murdering their wives. This was shortly before Egan murdered George Blake. Egan was sentenced to death but was reprieved by the Supreme Court on a technicality. A new trial was called, but the case was now considerably weakened, leaving the press to conclude that 'the luck of the hangman's rope has saved his neck'.[80]

A number of American reports noted the desire of execution crowds to obtain pieces of the gallows as mementos. The gathering of souvenirs at public executions was common enough across Europe as well. In January 1879, at a double hanging in the semi-public space of Indianapolis jail yard, the sheriff employed 125 men to control the crowd. However, 'while they were still hanging the crowd of invited guests began cutting pieces from the scaffold, and as soon as they were cut down there was a scramble for pieces of the rope.'[81] In March 1902, preparations for the hanging of Charles Woodward, in Casper, Wyoming, included crowd-control measures, with the anticipation that 'memento fiends will be on hand in large numbers to secure a piece of the rope and pieces of the scaffold'.[82] Following the hanging of Richard Mkwayne in York, Pennsylvania, in 1908, 'the rope was hacked into bits for souvenirs. Some of these changed hands at as much as $2 a piece.'[83] There clearly was a trade in obtaining the rope for public display. In 1893, one journalist explained how hangmen often sold off their ropes, or pieces of it, to 'dime museum managers'. A month after the execution, in November 1887, of the four Haymarket 'anarchists', condemned to death after dynamite was thrown at police during a labour demonstration in Chicago, several dime museums around the country exhibited uncut ropes that were purported to be those that hanged the men. This claim was made, despite the ropes having been cut into pieces just after the hanging and sold as per usual.[84]

The lynching of African-Americans led to the horrific free-for-all collection of body parts that has been described in terms of ritualistic sacrificial murder.[85] In 1906, a lynch mob took Nease Gillespie, John Gillespie and Jack Dillingham from Salisbury jail in North Carolina and hanged them. It was reported shortly after that 'thousands of people visited the scene of the lynching today. Many cut off fingers, ears, and other portions of the dead bodies, and carried them away as souvenirs.'[86] As well as hangings and lynchings, there are numerous reports during the American Civil War of soldiers taking body parts of their dead foes for trophies and souvenirs. Accusations of such activity were particularly levelled at Confederate soldiers, and despite the obvious propaganda aspects of such stories, the taking of body parts clearly happened. Local slaves reported to a Union army surgeon that Confederate soldiers had dug up soldiers' graves to make rings from the bones and drinking cups from skull tops.[87]

There was a tendency in the American press to assume that European-Americans were less 'superstitious' than their European cousins, and hence, perhaps, the regular reporting of hanging rope selling as part of the souvenir trade. How do we know, though, that these items were desired solely as macabre souvenirs? How many of these 'memento fiends' were actually acquiring them for magical and medical means? The Civil War was only a few decades on from the German reports of people taking fingers and other parts from the criminal corpses displayed on the wheel, and there were many ethnic German soldiers in the Civil War armies.[88] In England, the tradition of cramp bones kept in the pocket or under the pillow was widespread. While in most English folklore sources cramp bones were from the patella or knucklebone of a sheep or hare, one antiquarian noted in the early nineteenth century that he had 'heard that some of strong nerve, resolving to prevent the approach of so unwelcome an assailant as the cramp, have been known so temerarious as to wear the more potent spell of a human patella.'[89] We also need to bear in mind that mementos and souvenirs could accrue amulet-like power as they passed through different owners' hands over the years. Likewise, a piece of rope purchased for luck could, decades later, have no meaning other than as a historic curiosity, the motivation of its original owner completely lost to history.

Notes

1. See Spierenburg, *The Spectacle of Suffering*, ch. 3.
2. Steve Poole, '"For the benefit of example": Crime scene executions in England, 1720–1830', in Richard Ward (ed.) *A Global History of Execution and the Criminal Corpse* (Basingstoke, 2015), pp. 71–102. For examples in Scotland, see Rachel Bennett, *Capital Punishment and the Criminal Corpse in Scotland 1740 to 1834*, forthcoming.
3. Jorris Coolen, 'Places of justice and awe: the topography of gibbets and gallows in medieval and early modern north-western and Central Europe' *World Archaeology* 45 (2014) 767.
4. Evans, *Rituals of Retribution*, pp. 225–226.
5. *The Monthly Magazine*, 47 (1819) 509.
6. Sarah Tarlow, *Hung in Chains: The Golden and Ghoulish Age of the Gibbet in Britain* (Basingstoke, 2016); Zoe Dyndor, 'Gibbets in the Landscape: Locating the Criminal Corpse in Mid-Eighteenth Century England', in Ward (ed.), *Global History of Execution*, pp.102–126; Bennett, *Capital Punishment and the Criminal Corpse in Scotland*; Coolen, 'Places of justice and awe', 762–779; L. Meurkens, 'The Late Medieval/Early Modern Reuse of Prehistoric Barrows as Execution Sites in the Southern Part of the Netherlands', *Journal of Archaeology in the Low Countries* 2 (2010) 5–29; Nicola Whyte, 'The deviant dead in the Norfolk landscape', *Landscapes* 4 (2003), 24–39; Nicola Whyte, 'The after-life of barrows: prehistoric monuments in the Norfolk landscape', *Landscape History* 25 (2003) 5–16.
7. John Field, *A History of English Field Names* (1993), pp. 237–239; Johannes A. Mol, 'Gallows in Late Medieval Frisia', in Rolf H. Bremmer Jr., Stephen Laker and Oebele Vries (ed.), *Advances in Old Frisian Philology* (Amsterdam, 2007), pp. 276–277.
8. F. Hancock, *Wifela's Combe: A history of the Parish of Wiveliscombe* (Taunton, 1911), p. 253; *Taunton Courier*, 2 November 1921. See more generally, Owen Davies, *The Haunted: A Social History of Ghosts* (Basingstoke, 2007), pp. 51–54.
9. Ethel H. Rudkin, 'Lincolnshire Folklore', *Folklore* 44 (1933) 209.
10. Henry Swainson Cowper, *Hawkshead* (London, 1899), p. 326; Theresa M. Kelley, *Wordsworth's Revisionary Aesthetics* (Cambridge, 1988), p. 120; A. Craig Gibson, 'The Lakeland of Lancashire: Hawkshead Town, Church and School', *Transactions of the Historic Society of Lancashire and Cheshire* N.S., 5 (1865) 147.
11. *London Morning Penny Post*, 28 August 1751.
12. 'Spectre-Dogs', *Chambers' Book of Days* (Edinburgh, 1864), pp. 433–434.

13. Theo Brown, 'The Black Dog' *Folklore* 69 (1958) 176, 185; Davies, *The Haunted*, p. 37; Bob Trubshaw, *Explore Phantom Black Dogs* (Loughborough, 2005); Mark Norman, *Black Dog Folklore* (London, 2016).

14. Thomas Rudge, *The History of the County of Gloucester: Compressed* (Gloucester, 1803), Vol. 1, p. 239; Samuel Rudder, *A New History of Gloucestershire* (Cirencester, 1779), p. 606.

15. Thomas Rossell Potter, *The History and Antiquities of Charnwood Forest* (London, 1842), p. 179.

16. 'Hangman's Stone', *Manchester Weekly Times*, 27 May 1898; O.G.S. Crawford, 'Hangman's Stones', *Notes & Queries* 12 Series, xi (1922) 50-52.

17. Malcolm Jones 'The *Hangman's Stone* and the Unwonted Fruit: Two Emblems of Folkloric Origin', *Emblematica*, 5 (1991), 287–299; Arnoud S.Q. Visser, *Johannes Sambucus and the Learned Image: The Use of the Emblem in Late-Renaissance Humanism* (Leiden, 2005), p. xxiv.

18. Geoffrey Whitney, *A Choice of Emblemes* (Leiden, 1586), p. 41.

19. Peter King, *Crime, Justice, and Discretion in England, 1740–1820* (Oxford, 2000), p. 314, n. 51. See also, John Rule, 'The Manifold Causes of Rural Crime: Sheep-Stealing in England, c. 1740–1840', in John Rule (ed.), *Outside the Law* (Exeter, 1982), pp. 102–129.

20. Thomas Fuller, *The History of the Worthies of England* (London, 1662), p. 247.

21. Jacqueline Simpson, 'Seeking the Lore of the Land', *Folklore* 119 (2008) 133.

22. Alexandra Walsham, *Providence in Early Modern England* (Oxford, 1999), pp. 97–98; Alexandra Walsham, *The Reformation of the Landscape: Religion, Identity, and Memory in Early Modern Britain and Ireland* (Oxford, 2011), p. 499.

23. Charles Brewster Randolph, 'The Mandragora of the Ancients in Folk-Lore and Medicine', *Proceedings of the American Academy of Arts and Science* 40 (1905) 487–532; R.K. Harrison, 'The Mandrake and the Ancient World', *The Evangelical Quarterly* 28 (1956) 87–92; Jeannine E. Talley, 'Runes, Mandrakes, and Gallows', in Gerald James Larson, C. Scott Littleton and Jaan Puhvel (eds), *Myth in Indo-European Antiquity* (Berkeley, 1974), pp. 157–168; Marie-Françoise Notz, 'La Mandragore merveilleuse', *Cahiers Ethnologiques* 14 (1992) 129–142; Anne Van Arsdall, Helmut W. Klug and Paul Blanz, 'The Mandrake Plant and its Legend: A New Perspective', in Peter Bierbaumer and Helmut W. Klug (eds), *Old Names—New Growth: Proceedings of the 2nd ASPNS Conference* (Frankfurt, 2009), pp. 285–347.

24. Van Arsdall, Klug and Blanz, 'The Mandrake Plant', p. 333.

25. Laurent Catelan, *Rare et curieux discours de la plante Mandragore* (Paris 1638), pp. 3–4.
26. Randolph, 'The Mandragora', 495.
27. Jennifer Evans, *Aphrodisiacs, Fertility and Medicine in Early Modern England* (Woodbridge 2014), p. 178.
28. John Gerard, *The Herball or Generall Historie of Plantes* (London, 1633), p. 351.
29. Thomas Browne, *Pseudoxia Epidemica* (London, 1658), Book 2, p. 107.
30. William R. Newman, 'The Homunculus and the Mandrake: Art Aiding Nature Versus Art Faking Nature', in Jessica Riskin (ed.), *Genesis Redux: Essays in the History and Philosophy of Artificial Life* (Chicago, 2007), pp. 123–128; William R. Newman, *Promethean Ambitions: Alchemy and the Quest to Perfect Nature* (…), pp. 208–217; Frederick J. Simoons, *Plants of Life, Plants of Death* (Wisconsin, 1998), pp. 121–127.
31. Harrington, *Faithful Executioner*, p. 209.
32. Johann Georg Keyssler, *Antiquitates selectae septentrionales et Celticae* (Hanover, 1720), pp. 507–509.
33. *Secrets merveilleux de la magie naturelle et cabalistique du Petit Albert* (Lyon, 1752), pp. 166–168.
34. Thierry Zarcone, 'The Myth of the Mandrake, the "Plant-Human"', *Diogenes* 52 (2005) 115–129.
35. Rob Craig, *It Came from 1957: A Critical Guide to the Year's Science Fiction, Fantasy and Horror Films* (Jefferson, 2013), pp. 51–54.
36. Ruth A. Firor, *Folkways in Thomas Hardy* (Philadelphia, 1931), pp. 114–115; Edward Lovett, *Magic in Modern London* (Croydon, 1925), p. 74; *Nottingham Evening Post*, 12 June 1920.
37. Van Gent, *Magic, Body and the Self*, pp. 155–156.
38. Thomas Browne, *Pseudodoxia epidemica, or, Enquiries into very many received tenents and commonly presumed truths* (London, 1646), p. 272; Francis Grose, *A provincial glossary, with a collection of local proverbs, and popular superstitions* (London, 1787), p. 57; Per Binde, *Bodies of Vital Matter: Notions of Life Force and Transcendence in Traditional Southern Italy* (Gothenburg, 1999), p. 126; Brand, *Observations*, p. 583; Jacob Grimm, *Teutonic Mythology*, trans. James Steven Stallybrass (Cambridge [1883], 2012), vol. 3, p. 1138.
39. Strack, *The Jew and Human Sacrifice*, p. 75.
40. Tausiet, *Urban Magic*, pp. 93, 94; Víctor Lis Quibén, *La medicina popular en Galicia* (Pontevedra, 1949), p. 273.
41. Binde, *Bodies of Vital Matter*, p. 126; W.L. Hildburgh, 'Notes on some Flemish Amulets and Beliefs', *Folklore* 19 (1908) 204.
42. Harrington, *Faithful Executioner*, p. 187.
43. Brand, *Observations*, p. 583.

44. Ella M. Leather, 'Notes on English Folklore: Herefordshire' *Folklore* 27 (1916) 415.

45. Frederick Printer Miller, *Saint Pancras Past and Present: Being Historical, Traditional, and General Notes of the Parish*, etc. (London, 1874), p. 42; *Yorkshire Gazette*, 13 September 1845; *South Australian Register*, 11 April 1846.

46. *Stirling Observer*, 3 July 1856.

47. Davies, *Murder, Magic, Madness*, p. 162.

48. *Georgetown Gazette*, reprinted in the *Bradford Daily Telegraph*, 28 June 1871.

49. *Illustrated Police News*, 3 April 1886.

50. *Neue Freie Presse*, 13 February 1881; *Welt Blatt*, 16 February 1881.

51. Harald Seyrl, 'Todesstrage in Österreich (1876–1950)', in Maria Loredana Idomir, Matthias Keuschnigg, Michael Platzer (eds), *Vienna Conference on the Abolition of the Death Penalty November 2–13, 2011: Working Together towards the Universal Abolition of the Death Penalty* (Vienna, 2011), pp. 28–29; Anna Ehrlich, *Hexen - Mörder - Henker: Die Kriminalgeschichte Österreichs* (Vienna, 2006).

52. *Illustrated Police News*, 3 April 1886.

53. *London Telegraph*, reprinted in *Derby Daily Telegraph*, 30 March 1880;William Dixon Bancroft, *McKinley, Garfield, Lincoln; their lives—their deeds—their deaths—with a record of notable assassinations and a history of anarchy* (Chicago, 1901), p. 511.

54. *Western Daily Press*, 2 May 1881.

55. *The Near East*, 9 (1915) 406.

56. Widely reported in the press, though not always accurately. See, for example, *San Antonio Light*, 9 August 1925; *Pittsburgh Press*, 18 October 1932.

57. *Cornell Daily Sun*, 26 September 1924.

58. *Western Daily Press*, 15 August 1911.

59. *Colonist*, 15 August 1912.

60. Jacques-Barthélemy Salgues, *Des erreurs et des préjugés répendus dans la société* (Paris, 1910), p. 404.

61. Camille de Mensignac, *Notice sur les superstitions, dictons, proverbes, devinettes et chansons populaires du département de la Gironde* (1888), p. 130; Paul Sebillot, 'Les Pendus', *Revue des Traditions Populaires* 5 (1890) 592–594.

62. See, for example, *Western Gazette*, 24 April 1874.

63. *Gloucester Citizen*, 11 October 1883.

64. *Illustrated Police News*, 16 May 1885.

65. Nordiska Museet Arkivet, Botemedel 3, 23499, 23503, 28905, 26013.

66. *Gloucester Echo*, 7 November 1923.

67. Anne Plumptre, *A Narrative of Three Years' Residence in France* (London, 1810), vol. 3, p. 233.
68. *North Wales Times*, 2 April 1898; *South Wales Echo*, 21 June 1899.
69. *Portsmouth Evening News*, 24 May 1883.
70. Davies, *Murder, Magic, Madness*, p. 153; *Illustrated Life and Career of William Palmer* (London, 1856), p. 113.
71. *New Castle News*, 28 June 1911; Chris Woodyard, 'Enough Rope: The Hangman's Rope in the Press', http://hauntedohiobooks.com/news/enough-rope-the-hangmans-rope-in-the-press/.
72. *Trenton Evening Times*, 12 March 1909. Govogovitch's death sentence was commuted on a technicality.
73. Hand, *Magical Medicine*, pp. 71–72, 76.
74. Stuart Banner, *Death Penalty: An American History* (Cambridge, MA, 2002), pp. 154–156.
75. Annulla Linders, 'The Execution Spectacle and State Legitimacy: The Changing Nature of the American Execution Audience, 1833–1937', *Law and Society Review* 36 (2002), 607–656, esp. 627.
76. https://unitproj.library.ucla.edu/dlib/folkmed/index.html.
77. Reprinted in the *Wellsville Allegany County Democrat*, 14 June 1882.
78. *Boston Journal*, 23 June 1870.
79. *Rockingham Register*, 5 January 1888.
80. *New York Times*, reprinted in the *Dundee Courier*, 24 December 1885.
81. *New York Herald*, 30 January 1879.
82. *Denver Post*, 23 March 1902.
83. *Trenton Evening Times*, 12 March 1909.
84. *Reno Evening Gazette*, 12 July 1893.
85. See Amy Louise Wood, *Lynching and Spectacle: Witnessing Racial Violence in America, 1890–1940* (2011); Orlando Patterson, 'Rituals of Blood: Sacrificial Murders in the Postbellum South', *Journal of Blacks in Higher Education* 23 (1999) 123–127.
86. *New York Times*, 8 August 1906.
87. Simon Harrison, 'Bones in the rebel lady's boudoir: Ethnology, race and trophy-hunting in the American Civil War', *Journal of Material Culture* 15 (2010) 385–401; Simon Harrison, *Dark Trophies: Hunting and the Enemy Body in Modern War* (New York, 2014), pp. 93–117.
88. See, for example, Walter D. Kamphoefner and Wolfgang Helbich (ed.), *Germans in the Civil War: The Letters They Wrote Home*, trans. Susan Carter Vogel (Chapel Hill, 2006).
89. Roud, *Superstitions*, p. 119; Edward Moor, *Suffolk Words and Phrases: Or, An Attempt to Collect the Lingual Localisms of that County* (London, 1823), p. 90.

Lingering Influences

Abstract The physical influence of the executed criminal could live on long beyond execution, post-mortem spectacle and burial. This chapter explores how different cultures viewed and dealt with the spirits of executed criminals, and carried out a range of preventative post-mortem practices to ensure the dead did not come back to terrorise the living, whether in the guise of vampires or ghosts. It then considers why spiritualists in the nineteenth century actively sought out communication with the criminal dead. The curious American tradition and influence of Hangman Friday is also considered.

Keywords Hangman Friday · Living dead · Ghosts · Hauntings Spiritualism

The physical influence of the executed criminal could live on long beyond execution, post-mortem spectacle and burial. The resort to the relics of Christian martyrs for help and healing is an obvious example. The Catholic Church has long promoted the practice of touching or praying before the bones and associated objects of the saints. Many were attributed to biblical figures or the early church fathers. In south-eastern Italy, for instance, especially in the regions of Abruzzo and Puglia, epilepsy was known as the sickness of St. Donato, a beheaded martyr who

© The Author(s) 2017
O. Davies and F. Matteoni, *Executing Magic in the Modern Era*,
Palgrave Historical Studies in the Criminal Corpse and its Afterlife,
DOI 10.1007/978-3-319-59519-1_5

was invoked for healing and protection.[1] However, some popular relics belonged to more recent centuries, such as Catholic priests executed for their faith in Protestant territories. The most well-known example from England is Edmund Arrowsmith, a Jesuit executed in 1628. His hand was cut off and preserved, and into the nineteenth century Catholics and Protestants alike sought out its stroke as a miraculous cure-all. But the focus in this chapter, in keeping with the rest of the book, is with executed common criminals. We have seen how pieces of the criminal corpse and the hanging rope acted as talismans and charms, while sometimes, as with Mary Bateman, body parts retained their identity decades after, rather like saints' relics. Yet it was the spiritual, and not the corporal, remains that most often constituted the long-term, continued relationship between the executed and the living, an afterlife in which, freed from the criminal corpse, the criminal spirit could trouble and terrorise, but also atone through negotiation with the living. But let us start with a societal haunting of a most unusual kind: the malign shadow that hangings cast over the working week in America.

Hangman's Day

In nineteenth-century America, there was a widespread belief that Friday was an unlucky time to start a new piece of work or to embark on any ventures, because executions regularly took place on that day. In sympathetic association with the fate of the criminal on the gallows, any work begun would never be finished. Friday is obviously widely associated with the crucifixion of Christ, and the notion of it being unlucky was quite widespread. In England, there were folk beliefs (unrelated to executions) about it being ill-starred to be born on a Friday or to turn a mattress on that day, for example. There was a strong belief amongst fisherman that it was an unlucky day to set out for a catch.[2] However, the notion of Hangman's Friday is distinctly American, due to numerous states making Friday the customary time of the week for executions. In colonial eighteenth-century America, the English Murder Act of 1752 dictated that all convicted murderers had to be hanged within 48 h of sentencing, though Sundays had to be avoided. In the early Republic, Thomas Jefferson proposed, likewise, that the condemned be executed on the second day following conviction.[3] So how the Friday tradition became so engrained in American capital punishment custom is intriguing.

There are numerous references in early twentieth-century American folklore sources about Hangman's Friday, such as it was an inauspicious moment to travel or pick cotton.[4] The associations were not universally negative. In North Carolina, it was noted that Hangman's Day was a good time to plant pendulous fruits or vegetables, such as grapes and beans: the association is obvious.[5] But, basically, any other activity should not be commenced. The belief was clearly already pervasive by mid-century. In 1840, the Kentucky politician and lawyer, Henry Clay, wrote in a letter to the Virginian politician, Benjamin Watkins Leigh, 'I had thought of leaving Washn. For Richmond on Thursday, but now, I will not go until Friday, altho' I don't like that hangman's day.'[6] At the Massachusetts Horticultural Festival in 1848, one of the speakers observed how during his travels across the country:

> I have known housewives, otherwise most sagacious and sensible persons, utterly refuse to put their cloth in the loom, or their quilt in the frame, on a Friday; though everything should be ready and waiting to go ahead; as for putting their poultry on the goodly and important work of incubation on hangman's day; why Mr. President, it would be accounted nothing short of 'flat burglary' to think of such a thing![7]

It is possible that execution practices during the Civil War further heightened the tradition. Union executions for desertion were commonly held on Fridays, for example. In 1864, a Lieutenant in the 50th New Jersey Engineers wrote, 'Every Friday some poor victims has [sic] to pay the penalty of the law with his life. A week ago Friday I saw three hanged. Last Friday one was hanged, but I did not go to see him. Today there were two shot.'[8]

In April 1861, the California press commended a judge for setting an execution on a Tuesday. Unaware that the custom appears to have had its origins in Republican America, it opined that 'he has discarded one of the last relics of medieval barbarism … he has obliterated "hangman's day" from the calendar.'[9] Andrew Curtin, the Governor of Pennsylvania during the Civil War, apparently deliberately ordered that hangings take place only on Tuesdays, Wednesdays and Thursdays, in order to combat the belief in Hangman's Friday. The press welcomed his efforts to vanquish 'a relic of barbaric ages—the sooner forgotten, the better'. Governor John W. Geary, who replaced Curtin in 1867, was also praised for continuing the policy.[10] The Governor of Ohio's adoption of the same strategy that year was praised by the *Sullivan Democrat*, which noted that the attempt to eradicate the 'superstition' would be approved generally, not least because Hangman's Friday was disrespectful from a Christian point of view.[11] The idea that Hangman's Day was a retardant to economic development was also expressed. An editorial in the *Sullivan Democrat* complained that 'some people are so prejudiced against the fifth day in the week, that they would not, for a moment, entertain the idea of changing location on that day, or open a new business, although not doing so would be a pecuniary damage to them'. The *Augusta Chronicle* similarly argued, 'in this busy age we have no day that we can afford to brand with a bar sinister'. These comments echoed some of the economic arguments put forward against public execution as well. A report written for the Pennsylvania legislature by Jacob Cassat, in the 1820s, made the point that, because of the large crowds attracted to hangings on a Friday afternoon, labour and commerce were disrupted, thereby having a detrimental material impact on the 'public interest'.[12]

However, despite Ohio and Pennsylvania taking a state-wide stance, elsewhere the practice persisted, because it was ultimately left to the predilections of judges and governors to decide hanging days. The issue arose periodically when judges shied from tradition. When the murderer, Menken, was executed on a Wednesday in Elmira, New York, in 1899, it was suggested in the press that the judge had made the decision to 'weaken the prejudice' against Fridays.[13] When, in 1901, New York State decided to execute the assassin of President McKinley on Monday 28 October, the *Augusta Chronicle* used the opportunity to request that the Georgia Superior Court Judges likewise cease Friday executions, in order to undermine the 'superstition' of Hangman's Day. 'The idea cannot be erased in a short time,' it admitted, 'but there is no reason why the good work should not be inaugurated at once.'[14] Despite the clear concern over the influence of Hangman's Day, there was no concerted federal attempt to reform the custom. The efforts were led, instead, by a group of concerned members of the public.

Founded in New York in 1882, the 13 Club aimed to vanquish the 'slimy coils' of that 'hydra-headed monster', superstition, wherever it was to be found in America. As the minutes of the club for 1895 record, it considered its campaign against Friday hangings one of its greatest achievements. Its first archivist, Marvin R. Clark, led the fight. He wrote letters to judges and governors whenever a hanging was announced, requesting that they sentence the condemned to death on any other day than Friday. According to the minutes, several judges acceded to his request, and the club had in its archive a letter from David B. Hill, former Governor of New York between 1885 and 1891, which related that Hill had reprieved the execution of one murderer solely to avoid a Friday execution and so break down the superstition.[15] However, despite the hubris, the 13 Club's influence was limited. The tradition was too engrained. Friday executions continued to be the norm in Texas during the 1890s and into the early twentieth century. In 1905, the *Fort Wayne Sentinel* exclaimed, with apparent concern, 'Hangman's Day is to be Violated: Execution at Michigan City in June will not take place on Friday'. It was only the second time in the history of Indiana's state prison that this had happened. California was still holding to the Friday execution tradition during the mid-twentieth century.[16]

GHOSTS OF EXECUTED CRIMINALS

The ghosts of executed murderers were far less numerous, it would appear, than their victims. The long-held concept of purposeful ghosts, which lingered on earth, suffered in Purgatory, or returned from the heavens to correct the injustices that befell them in life, provided little scope for executed murderers to do anything useful amongst the living. There was no point in returning to show remorse to their victims— that was an afterlife matter. The ghosts of murderers generally served the function of spiritual memorials to the violent termination of life, as reminders of the fate that awaited those who strayed into crime. So, as noted in the previous chapter, we find the ghosts of the executed lingering at the spot where they were executed and gibbeted, long after all visible signs of punishment had been expunged from the landscape, with only a place name, perhaps, otherwise to memorialise the execution. Guy Beiner's work on the folk commemoration of the Irish Rebellion of 1798 shows how communities generated the ghosts of executed rebels as memorials to national grievance, as well as to personal and local injustice. The gallows sites where the British used to execute the rebels subsequently became *foci* for legends of fairy and ghostly activity that still circulated in popular culture into the twentieth century.[17] Where national identity was not involved, the spirits of the crime scene execution and gibbet were bound to fade into vague memory several generations after such public penal practices ended. In 1926, it was reported that the site of the former gallows near Marine Terrace, Aberystwyth, had once been haunted by disreputable ghosts, for instance, though a modern hotel that stood on the spot was not plagued at all.[18]

As jails and prisons became the main locations for executions, it is not surprising to find increasing reports of ghosts haunting the cells of the condemned and those awaiting trial, rather than spooking weary travellers halting at rural crossroads at night. In 1845, for instance, a young man convicted at Inverness assizes for issuing base coin requested to be removed from his cell in Tain jail after seeing the ghost of a man pacing back and forth muttering, 'Do it, do it, do it.'[19] American jails seemed to have been particularly prone to such hauntings. In 1868, Chicago jail was reported to be haunted by the ghost of one Fleming, who had been hanged there 2 years before. Cries of distress were heard at night, while two guards reported hearing the agonised words, 'Oh dear!' emanating from the jail vault. An African-American inmate, named William Jones, said he saw the apparition of a man one night in his cell with a strap around his neck.[20] Seven years later, Bergen County jail, Hackensack, New Jersey, was haunted by the ghost of John W. Avery, a young man executed there on 28 June 1872. Prisoners saw a strange bluish light and felt cold rushes of air as a shadowy figure stalked the cells at night, on one occasion opening Avery's old cell door. The bedclothes of a German inmate who occupied Avery's former cell were pulled off one night by the spirit.[21] In Birmingham jail, Alabama, in 1888, inmates swore to looking out over the courtyard and seeing the ghost of George Williams. He had been hanged in the yard a few months before, for killing a fellow convict. The authorities had left the scaffold standing and the rope dangling as a warning. Inmates repeatedly saw his ghost ascend the scaffold, adjust the rope around his neck, and drop through the trap. These sightings led to a great religious awakening in the jail, with regular psalm singing and prayer meetings.[22] In April 1908, the inmates of the county jail in Asheville, North Carolina, were deeply disturbed by the supposed ghost of a recently executed African-American. Days after the hanging, they nightly heard the loud crash of the scaffold trap, and some inmates and guards said they saw the man's ghost swinging in front of the iron gratings of the cells. The inmates prayed, but to no avail, and so they presented a petition, requesting that the authorities do something about the haunting. It was promised they would be removed to another jail.[23]

With these jail ghosts, both places and people were being haunted. In terms of place, it was not only the inmates who were sensitive to the possible spectral visitation of executed criminals. As noted above, jailers sometimes reported ghosts. Soon after the execution of Martha Alden in Norwich, in 1807, stories circulated that her ghost 'walked' the environs of the prison on Castle Hill. Several drunken men who attempted to lay the spirit were seized by the jailer and detained in prison.[24] In January 1883, a soldier was sentenced to 12 months' imprisonment for leaving his post outside Galway Jail. He pleaded that he had only left because he had been frightened away by the ghost of Myles Joyce, who was one of three men executed by William Marwood the previous month. On the scaffold, Joyce had professed his innocence of the crime.[25] And, of course, ghost belief had always been fruitful for hoaxers and fraudsters, so sometimes there really were sounds and sights that were meant to be taken as hauntings. In 1749, several notorious burglars in New Gaol, Southwark, began knocking on the walls of their cells for several nights, claiming that they were the rappings of several executed criminals who haunted the cells. The whole thing was a ruse to distract the jailers from the escape they were planning through a breach in the jail wall.[26]

In terms of people, the guilt-ridden murderer metaphorically haunted by the ghost of his victim was a widespread notion, but with these jail hauntings we have the criminal dead as spiritual reminders of the potential fate awaiting their former convict community. Some condemned men, such as John Avery, vowed to return and haunt their jailers and inmates after their execution. Sometimes, this spectral vengeance was directed at those who were instrumental in their conviction. When, in September 1894, the American robber and murderer, Thomas Dennis, returned to his cell after being sentenced to death, he wrote on his cell wall: 'I die innocent of this foul charge and my ghost will forever haunt those who made it.' He then hanged himself with his braces. The accusation that so enraged him was that he had once been a New York policeman.[27]

There were two individuals in particular that the spirits of the executed might have had cause to haunt: the executioner and the anatomist. The latter profession seems to have been passed over, however. Perhaps this was because the act of dissecting the corpse destroyed the individual identity, the completeness of the buried, soulless body, which was necessary for ghosts to be generated in the human mind and culture. Apart from the familiar stories of headless ghosts, I have not come across any dismembered ghosts. The anatomist, of course, merely handled the bodies of those already deceased at the hands of the executioner.

If guilt created metaphorical and 'real' ghosts in popular culture, the anatomist was largely absolved, but the executioner was surely ripe for haunting. The German-born American hangman and jail guard, George Maledon (1830–1911), was apparently once asked by a lady whether he was haunted by the ghosts of the eighty or so men he had hanged. He replied, 'I have never hanged a man who came back to have the job done over.' Legend also has it that when Deputy US Marshal Heck Thomas asked the same question about ghosts, Maledon said, 'I reckon I hung them too'.[28] During the early 1890s, he earned an income giving lecture tours as the 'Prince of Hangmen', displaying a collection of his ropes, bits of the scaffold and photographs of his hangings. However, 5 years after his retirement in 1894, it was reported widely in the American press that Maledon was unable to sleep and was badly tormented by what he thought were the spirits of his victims. Reports said that he kept the oil lights burning all night in his farmhouse in Washington County, Arkansas, to keep them at bay. 'Take them away! Take them away!' he apparently moaned in his sleep: 'the old man's watchers know that he is dreaming of his ghosts.'[29] Maledon was annoyed by these reports and denied he suffered in this way, but the story had much greater public traction than his denials. When, in 1903, he attempted to come out of retirement and officiate over a hanging at South McAlester, one news-paper commented that his renewed enthusiasm for the job was 'one of the most convincing evidences that the story of his ghost is emphati-cally a ghost story'.[30] As this story suggests, there was a desire to want to believe that the executioner was racked with guilt and remorse, as a means of assuaging and displacing societal guilt for state-sanctioned kill-ing. So, folklore generated the appropriate ghosts to reinforce this cul-tural impulse. A gallows site in County Mayo used to execute rebels in the 1798 uprising was, for example, still thought to be haunted by a guilt-ridden executioner in the 1930s.[31]

SUICIDE GHOSTS

Across Catholic, Orthodox and Protestant Europe, suicides constituted a difficult, feared category of dead criminal. Suicide was an act of self-execution, but there could be no public redemption or remorse on the scaffold. Self-murder was the most heinous of sinful crimes. Killers did not damn the souls of their victims, but self-murderers deliberately damned their own souls. This theological position occasionally drove the suicidal to the act of murder in order to be executed by the state, thereby achieving their desire to end their lives and still leave open a possible path to salvation.[32] It is hardly surprising, therefore, that suicides, like executed criminals, were thought to leave particularly troublesome or persistent spirits. This led, in turn, to the development of a range of ritual practices to dispose of the polluted corpses as, by law, they could not be buried in consecrated ground. In medieval Norway, France and Italy, for instance, burial on river banks and the tidal zone on seashores was practised—areas away from the community, the water perhaps purifying but also acting as a boundary, drawing the troubled spirit away from the living. The staking or weighing down of corpses in bogs, woods and other marginal landscapes is also well attested from medieval archaeology.[33] Such practices continued into the early modern period. In Augsburg and Metz, suicide corpses were nailed in barrels and floated down river, thereby cleansing the local populace, and generously passing on potential troublesome spirits to other communities. The same practice remained on the Swedish statute books until 1736, as did the alternative of burning suicides in the woods. In Nuremberg, corpses were sometimes burned at crossroads.[34] The law in England stated that suicide corpses should be interred in or by the highway, usually at crossroads, sometimes with a wooden stake driven through the body, though this was not an action specified by coroners' juries.[35] Road improvement and widening schemes during the late nineteenth and early twentieth centuries occasionally turned up examples. It was reported in 1899, for instance, that the widening of a crossroads near Birmingham had unearthed the skeleton of man with a stake through his sternum.[36] On the Continent, it was often the job of the executioner to deal with the burial of such tainted corpses in these ways—and it was a profitable business. One early modern Munich executioner went to court against the city's knackers to protect his monopoly.[37]

It would be highly misleading to suggest that most suicide corpses in the early modern period were subject to such rites of desecration and post-mortem punishment. It was probably mostly applied in aggravated cases, such as murderers, or other criminals who committed suicide to avoid trial and state execution. There was, furthermore, growing medical and moral recognition of the relationship between mental illness and suicide, which led to the more compassionate treatment of suicide corpses when it was determined by the coroner, or other authority, that the act of self-murder was committed whilst insane. In 1742, the Augsburg government decreed that all suicides be given a consecrated burial, apart from those who were convicted criminals who killed themselves. The city executioners were not happy with this further loss of revenue.[38] In England, there was a long tradition, carried on into the nineteenth century, of burying suicides and unbaptised babies on the shaded north side of the churchyard.[39] Even if suicides were not buried in consecrated land, by the nineteenth century, it was increasingly standard across much of Europe for suicide corpses to be buried quietly and without ceremony close by or alongside churchyard walls, as in Abruzzo, Italy.

Still, across Europe, sensational instances of post-mortem punishment continued at a local level into the modern era. An entry in the parish register of Hailuoto Island, Finland, for 1761 records: 'in this parish took place the deplorable event of farmer Henr. Pramila hanging himself; since it was a suicide, on 10/3 he was buried aside in the woods of Hanhis Hill by executioner Rönblad'. In November 1803, a dressmaker, named Lindemann, hanged himself in the town of Jüterbog, Saxony. By law, the local authorities should have informed the nearest anatomy school so that they could collect the corpse, but instead they ordered the local executioner to drag his body through the streets of the town and bury it under the gallows.[40] In England, a survey of East Anglian newspapers revealed 33 instances of roadside burials between 1764 and 1823. In six of the cases, the corpses were also transfixed by a stake.[41] This practice was ended by an Act of 1823, but the same law ensured that a ritual stigma remained, by stipulating that suicides would not receive Christian rites and that their burial could only 'take place between the Hours of Nine and Twelve at Night'. When this was repealed in 1882, the new statute erased the time limitations but still required burials to take place after sunset.[42] By the early twentieth century, this was being described as 'a barbarous survival' by one newspaper, while in 1902 the Reverend J. Trew of Batley, Lancashire, was booed by a crowd of 200 people after he refused to bury the corpse of a local mill-hand who drowned himself until after nine o'clock in the evening.[43] Similar restrictions were in place in the Swiss Canton of St. Gallen in the nineteenth century. Suicides were buried at night or at dawn in a secluded spot of the churchyard, with a possible private ceremony reserved for the relatives.[44]

The notion was quite widespread that the spirits of suicides, as with gibbeted criminals, manifested themselves in non-humanoid forms. In Valtellina, northern Italy, they appeared as big cats, whirlpools, and inanimate and animate shapes. Their wanderings did not stop until the day on which they should have died naturally.[45] In Normandy, France, the souls of suicides were thought to linger as a black dog, known as the *varou*. The same notion is known in England, too, with demonic saucer-eyed black dogs associated with the locations of violent deaths more generally. In 1851, it was reported that an old crossroads suicide burial near Boston, Lincolnshire, England, was haunted by a spirit, or 'tut', known as a shag foal. This was a variant of the black dog but in the form of a horse with saucer-like eyes.[46] The notion of the soul not being able to take on the form of its human vessel might reflect an echo of purgatorial punishment in popular eschatology.

The spirits of suicides were often associated with the spots where they drowned or hanged themselves. What concern us here, though, are the location and conception of spiritual disturbance associated with the burial and post-mortem treatment of suicide corpses. People had little influence over the locations where suicides took place, or where an execution occurred, but they could attempt to influence the corporeal source of potential troubled spirits. Because of the post-mortem treatment and burial of suicides, their spirits were less contained, more spread across the landscape, and more prone to wandering abroad. In central and eastern Europe, this sometimes manifested itself in concerns over vampires and the physical walking dead, and the practice of staking was just one of several judicial and folkloric means of revenant prevention.[47] In Russian folk religion, there was a strong link between suicides and demonic forces. It was the Devil who drove people to commit self-murder, so that he could then manipulate their spirits, riding and goading them for his pleasure.[48] In early modern Germany and Austria, popular concern led to several so-called 'suicide revolts', or 'cemetery revolts', where communities attempted to prevent the Christian burial of a suicide from taking place. It was thought that terrible thunderstorms would result if the full rites were given.[49]

This was not just an early modern problem. During the nineteenth century, popular fears led to resistance against more enlightened church policy and practice towards the burial of suicides. In Zuckenriet, St. Gallen, in 1827, villagers began to hear frightening noises and experience feelings of loathing and delirium shortly after a woman who committed suicide was buried along the wall of a churchyard. They dug up her corpse during the night and reburied it in a bog. A foreigner who had accepted to help with the woman's funeral was ostracised.[50] Fifteen years later, the *curé* of Champtoceaux, France, wrote to the bishop of Angers about his experience dealing with a parishioner who had recently committed suicide. She was a woman who had suffered from leprosy. The *curé* had tried to convince her family that she had been insane at the time of committing the act, so that he could bury her in consecrated ground and give her a proper Catholic burial service. However, they persevered in saying she was sane, and therefore he desisted.[51] When Russia was hit by a series of poor harvests and famine in the late 1880s and early 1890s, there was a wave of grave desecrations, as it was thought that suicides that had been given a Christian burial were the cause of natural disasters. In 1892, for instance, a Russian priest was sentenced to penance in a monastery for allowing his parishioners to disinter a suicide corpse and dump it in the woods.[52]

SANCTIFYING THE EXECUTED SPIRIT

In some cultures, under certain conditions, the ghosts of the executed or untimely dead were not to be feared or avoided: they were to be reintegrated into society. Purgatory and the destiny of the untimely dead mingled synchronically in the representations of the souls of executed criminals. In Naples, this was manifested in the long tradition of the *anime pezzentelle*, or *capuzzelle*, the forgotten souls of those who died a violent death, or in misery and abandonment. In Catholic tradition, Purgatory was a reservoir of both useful and frightening souls and ghosts. This intermediate state might become visible in earthly locations: in fact, God sent the souls to specific places to cleanse themselves of their sins. They inhabited the landscape, and, as attested by traditions from central Italy, they could be seen at night or at midnight in the form of unloved animals, or as little flames and shooting stars. In Sardinia, it was believed that spotting or interrupting these souls during their penitential activity could bring about retaliation, ranging from mild to lethal punishments. In the Sardinian regions of Barbagia and Oristano, the inhabitants of Purgatory came to succour the living. In Oristano, this belief was especially linked to magical practices, involving strange rituals and midnight prayers under the scaffold, or close to the remains of executed criminals, in order to gain the favours of the beheaded souls. The souls of executed criminals aroused both horror and compassion in northern Italian regions, reappearing in devotional creeds. Peasants from Valtellina invoked them in their prayers up to the nineteenth century, while in

Veneto they were connected to the damned dead, the murdered and the suicides.[53] According to folklore, sixteenth-century Venetian witches employed them in their incantations, praying especially for a certain Antonio dal Pomo d'Oro, who had the double status of murderer and suicide, having hanged himself after killing his whole family.[54]

Italian executed criminals survived in the daily preoccupations of the living through beliefs and folklore about the afterlife, conceptualised in the destiny of their souls. According to a belief widespread in the town of Potenza, for instance, during the eve of the 2 November, the souls of the deceased were united with their skeletons, reacquiring their fleshy appearance to walk the streets and be reacquainted with relatives and friends. Among these souls, there were even those who had died a violent death and the *disgraziari* (disgraced people), that is executed criminals. The murdered exhibited their bleeding wounds, while the executed wore the hanging rope around their neck or bore the severed head in their hands. If the believers had a friendly disposition, they would have been advised of the coming of the *disgraziari* by the souls themselves, but it was not allowed to talk to these dead, otherwise they would disappear.[55]

The threshold of death implied the detachment of the individual from the body and the passage of the soul, which, through the public spectacle of the scaffold, was shared and ritualised by the whole community. Punishment and justice turned to piety and prayers to reintegrate the soul of the executed criminal within society: the freshly executed corpse represented the beginning of a purgatorial experience to reach heaven. Good examples come from southern Italy, especially Sicily, where the souls of repentant executed criminals were popularly sanctified. The Company of the Saintly Crucifix or *Bianchi* (the White Ones), from Palermo, took care of the condemned for 3 days and nights before the execution, providing religious comfort and accepting them as a member of the brotherhood. This operation helped criminals to cope with their approaching death, but it also placed them in a liminal zone, where the detachment of the soul from the body brought a social transformation from common individuals to the emblems of Christ's sufferings on the cross. Once dead, they were buried in the Church of the Saint of Beheaded Souls, under the protection of John the Baptist, the most famous beheaded saint. Here, they attracted collective concerns and devotional practices. People attached to them either their personal hopes for redemption or more practical daily requests.[56]

Many fraternities of the dead were founded between 1350 and 1550 in towns and cities across Italy. They were formed to look after the spiritual needs of the condemned before and during their execution. However, as Michael P. Carroll has suggested, there is nothing in the sources to suggest that before 1600 these fraternities were also concerned with special devotion to the corpses of the executed. In other words, it was during the seventeenth century that interest in the purgatorial executed shifted from them being seen solely as troublesome spirits that needed to be assuaged and kept away, to souls that could serve a valued post-mortem function. This led to the formation of popular skeletal cults of the hanged and beheaded, with their physical remains becoming objects of devotion.[57] Often considered as a macabre medieval survival, it appears that the cults of the beheaded, and the like, were more recent urban and rural responses to shifts in popular attitudes to justice and Purgatory. Political and social issues flowed into religious ones and the criminal could become a popular hero in death. Exceptional in their guilt, and often outsiders as living beings, executed criminals received care and compassion after death, embodying the fate of the repenting robber on the cross.[58] One example is the *anime mpilluse*, from Messina, who took their name from the 70-year-old Andrea Belluso, a generous merchant hanged by the Spanish in 1679 as a rebel, but seen as an innocent man by the population, who made a saint of him. Later, the same devotion was reserved for Francesco Frusteri, who was beheaded at Paceco, near Trapani, in 1817, for having killed his own mother. Because he showed exceptionally deep repentance, he became a popular, though not officially recognised, saint, to whom miracles were attributed. People visited his grave to request his intercession and help with problems.[59] So the souls responded to the needs of the weak. In Palermo, they could occasionally appear during night-time, walking in white garments near the River Oreto, not far from their church, and speaking in muttered words, to advise, warn, and help people with their problems.[60] Not all spirits of the executed were benign or helpful, though. The story of Antonio Cilizza, known as Tulé, is indicative. He was a ferocious criminal, hanged on the plain of Terravecchia in 1747. After the execution, his blood gushed off the platform, preventing his soul from reaching the otherworldly abode to which it was destined. As a consequence, the soul of Tulé continued to prowl the streets, making noises and preventing muskets from firing, frightening the population for more than a century after his death.[61]

As the case of Frusteri shows, new popular cults of the beheaded souls were being formed into the nineteenth century. Up to the early twentieth century, during the night between 28 and 29 August, women went to the Church of Saint John the Beheaded on Mount Andria, outside the town wall. They entered though a lateral door to leave an offering and pray to the executed criminals to protect their sons from awful deaths.[62] The interior of one such church in Milan was described in 1865. Hanging on the wall:

> was a large painting that showed the members of this confraternity accompanying a condemned prisoner to the place of execution ... Two skeletons hung beside the painting, one on each side. One skeleton had a noose around its neck, while the other was holding its own skull in its hands ... Underneath were reliquaries containing sixteen skulls and one bony torso, placed there to collect alms in support of the executed criminals buried in this church over the course of three centuries.[63]

In practical terms, the souls were employed in several charms. 'Three hanged, three slain, three drowned', was the trinity to whom prayers for health and protection were recited in parts of southern Italy.[64] People asked them to intercede with God, but they also invoked them for revengeful purposes against enemies. Some attempted to extort favours from the souls, in exchange for their devotion; for others, love magic was the main concern. Women implored the souls of the executed to beat their lost lovers to return to them. In early modern Spain, gamblers sought to enhance their luck by trying to get the souls of those executed by hanging to intercede on their behalf. To that end, they would carry a length of the hanging ropes and pray for their souls.[65]

SPIRITUALISTS REACH OUT

The Italian Catholic traditions show that the spirits of some executed criminals were not always feared and avoided. With the advent of the spiritualist movement in the mid-nineteenth century, a new Protestant desire to communicate with them opened up as well. While the Catholic engagement with the executed was couched in ritual and intercessionary terms, spiritualist engagement was evangelical in its desire to reform the living dead, rather than seek their succour. But both faiths gave the opportunity for the executed to express repentance. So, the early

American clairvoyant and medium, Sarah Danskin, proactively sought out the spirits of the criminal dead in order to enable them to atone and progress in the spirit realm. Mediums such as Danskin were afterlife missionaries, enabling salvation for such 'dark' and 'undeveloped' spirits.[66]

In 1857, the Nottingham healer, seer and spiritualist medium, John G.H. Brown, published his purported spirit communications with a couple of notorious murderers, whom we have already encountered earlier in this book. One was the poisoner, William Palmer, executed a year earlier. The other, William Saville, was hanged in Nottingham, in 1844, for the murder of his wife and three children.[67] At least twelve people were crushed to death at Saville's execution as the huge crowd began to disperse, adding to the potential post-mortem weight of guilt on his ethereal shoulders. Brown used a crystal ball to contact angels and the denizens of the afterlife.[68] It was in this way he asked the Archangel Gabriel to ask whether Palmer had been guilty or not, to which Gabriel said he would command Palmer's spirit to appear and give a truthful statement. 'When the darkness of the vision cleared off,' wrote Brown, 'to my surprise the figure of a man appeared ... exhibiting no signs whatever of a spiritual appearance, which greatly astonished me. He held no scroll nor showed any symptoms of communication for some short time; till at length, *another figure* appeared, adorned with loose bright robes'. This divine figure gave the spirit of Palmer (for it was he!), a scroll, which was then opened and displayed in the ball, allowing Brown to make a copy.[69] It began, 'I am the spirit of William Palmer ...' and went on to confess to the murder of J.P. Cook and five other people by poisoning:

> On passing from life to immortality, in the manner publicly described, I experienced those pangs which I have since learnt others have described. I have also heard the yells, groans, and shrieks, beyond the darkness; suffered the taunts and reproaches of my murdered victims; and am now dwelling in the atmosphere, around and near the scenes of my worldly existence; experiencing the bitter reproaches from the thoughts of those who are living ... Oh horrible! Horrible! Wretched misery! And terrible but mysterious immortality! I must now leave you. My victims haunt me!

Palmer's supposed revelation from the spirit world made no reference to him being in hell for his sins. The spiritualist movement, in general, did not believe in the old concept of the infernal regions. There was a

structure and progression through the afterlife, though most spiritualist texts focused on the pleasant middle realms, where the more recent,
decent, dead dwelled, and from where most spirit visitors made their
spectral manifestations to the living. Still, the executed obviously could
not be expected to go straight to the middle realm, and there was an
appropriate level of transitional, self-imposed punishment and discom-
fort for them. Therefore, the lower spirit regions were usually described
in terms of 'darkness' with a 'Hell-lite' touch of torment and torture.[70]
Executed criminals were given a second chance, with or without the aid
of mediums, to redeem themselves, so Harvard lecturer, John White
Webster, executed in 1850 for the murder of Dr. George Parkman,
apparently began his afterlife journey through the realms by first trying
to find his victim. It was in vain at first, and he spent his time alone in
darkness. 'At last, however, I met with Dr. Parkman, and obtained his
forgiveness. I cannot tell you the weight which seemed removed by it. I
then knelt, and with all my soul sought pardon of my Maker ...now I am
in a somewhat more hopeful state'.[71]

There ran through the nineteenth-century spiritualist movement a
strong vein of support for the abolition of capital punishment.[72] This
was, in part, drawn from the influence of liberal, universalist and evangelical theologies that promoted redemption, universal salvation, passion
and forgiveness.

In spiritualist terms, execution did not act as a punishment, in the
sense that it extinguished sentient life. Indeed, it prematurely transferred
the dark souls of criminals to a more advanced state of being. In pragmatic terms, this also meant that the gallows, unnecessarily and perni-
ciously, let loose a host of very bad spirits. Writing in 1856, Newton
Crosland, husband of spiritualist writer and novelist, Camilla Dufour
Crosland, stated, for instance, that 'a believer in Spirit-manifestations
cannot consistently approve of capital punishments', for the wicked
souls of executed murderers could wreak woe and destruction more
fatally than their former bodies could perpetrate if they were still alive.[73]
Writing in the 1890s, the Irish-born American Theosophist, William
Q. Judge, explained that this was because the untimely dead were not
fully deceased during their time in the astral plain known as *kamaloka*
(a term borrowed from Buddhist belief). Ultimate death and spiritual
release could only occur once their natural period of life had expired,
'whether it be 1 month or 60 years'. It was these limbo-like astral
shells, rather than the actual souls of the dead, with whom spiritualist

mediums communicated, he believed. The experience of life in *kamaloka* depended on the moral status of the individual. 'Executed criminals are in general thrown out of life full of hate and revenge, smarting under a penalty they do not admit the justice of,' Judge affirmed. 'They are ever rehearsing in *kamaloka* their crime, their trial, their execution, and their revenge.' Therefore, when a medium provided these 'spooks' with an opportunity to communicate with those on earth, they injected into living persons 'deplorable images of crimes committed and also the picture of the execution and all the accompanying cures and wishes for revenge'. And so in those countries that practised capital punishment, 'crimes and new ideas of crimes are wilfully propagated every day'.[74] The dangers of communicating with the executed were revealed in the early days of modern spiritualism during the mid-nineteenth century.

In November 1849, the body of a German peddler, named Nathan Adler, was found by a roadside about sixteen miles south of Auburn, New York. Albert Baham, one of three brothers involved in the crime, was sentenced to death, and cursed and railed at those whose evidence led to his execution. A spiritualist group employed a young female medium to use her clairvoyance to witness Baham's execution, in order to 'observe the separation of body and spirit, and the manifestation of the latter'. The experience was so profound that the medium fainted as Baham's corpse fell through the trap, and from then on Baham's spirit regularly communicated with her, she said. He cursed his enemies, vowing revenge upon them and threatening to cause the deaths of several Auburn residents who had crossed him. As time went on, Baham's spirit began to take physical control of the medium, causing her to beat her arms against her chair until they were black and blue. Then she began to experience the sensation of a rope being drawn around her neck, while Baham said he wished to strangle her to death. Physicians were consulted to sooth her bodily torment, and magnetisers attempted to rid her of the malignant spiritual influence. All to no avail. She was finally rid of Baham's possession by another medium, who called upon the aid of St. Paul.[75] The cautious spiritualist, Eliab Capron, who knew the medium concerned, first printed an account of the case in 1855, observing the dangers of communicating with 'coarse and undeveloped' characters such as Baham.

Some thought the case was one of pure diabolism. The spiritualist critic, William R. Gordon, reprinted the Baham possession account in order prove that spiritualism was no more than a 'revival of heathenism without the pomp of Popery', a vehicle for the Devil to corrupt Christian society. Mediums were being tricked into believing they conversed with the criminal dead by evil spirits.[76] This did not stop mediums, though. The poet, radical politician and spiritualist, Gerald Massey (1828–1907), attracted some notoriety with his spiritualist dabblings in the sensational Franz Müller murder trial of 1864—best known as the first British railway carriage homicide. At a séance, held while the trial was ongoing, Massey claimed to have received a spirit communication from the victim, Thomas Briggs, claiming 'Müller not guilty; robbery, not murder'. Shortly after Müller was hanged, his spirit contacted Massey through a female medium, and 'purported to come and thank me in trying to save his poor neck'. Massey also reported in a lecture that one of his family's former homes had been haunted by the restless spirit of a child murderer. As in the Baham case, the female medium suffered terribly from being possessed by the murderer's spirit.[77]

Finally, and even more sensationally, spiritualism offered the prospect of getting a unique personal account of the experience of being executed—from the executed themselves. Here is what that the spirit of William Saville apparently told the medium, John G.H. Brown:

I ascended with firmness, accompanied by the officers and other functionaries. I viewed with horror the immense assemblage which had collected to witness the last penalty which the law could inflict upon me for my crimes, and recognised several persons of both sexes in the crowd. The sensation of the operation performed by the executioner in adjusting the fatal rope can be described by none but those who have experienced it. At length I found myself blufted [blinkered] from the light of the world, and, after the usual words from the functionary on such occasions, I felt the horrible sensation of tottering and trembling on the verge between life and immortality – a sharp and momentous click which ran through my frame with indescribable horror, and the next moment I felt myself drop for several feet. At the same instant, indescribable pain convulsed my whole frame, and a noise as of many heavy carriages passing over the paved streets filled my ears, and my heart felt as if seized by a hand of ice, which forbad its functions. My limbs then appeared to be set fast; a death-like faintness came over me, and the same moment I experienced the sensation as of a sleeping vision, and all pain, all cares, and all troubles left me. My eyes then appeared to open, and I felt conscious of what had passed, heard the screams which ascended from the crowd below, and felt to secretly smile at their belief of my being dead.[78]

NOTES

1. Mariano Cipriani, 'Contributo allo studio dei vecchi appellati-vi agiogra-fici del Mal Caduco', *Rivista di storia della medicina* X, 1, (1966) 96; Adriano Puce, 'Il male di S. Donato nel Salento. Contributo psicologico-sociale', *La Ricerca Folklorica* 17, (1988) 43–59.
2. For example, Roud, *Penguin Guide to the Superstitions of Britain and Ireland*, pp. 197, 217.
3. John E. Ferling, *Setting the World Ablaze: Washington, Adams, Jefferson, and the American Revolution* (Oxford, 2000), p. 160.
4. Newman Ivey White and Wayland D. Hand (ed.), *The Frank C. Brown Collection of North Carolina Folklore* (Durham, NC, 1964), pp. 197–199.
5. N.C. Hoke, 'Folk-Custom and Folk-Belief in North Carolina', *Journal of American Folklore* 5 (1892) 113.
6. Melba Porter Hay (ed.), *The Papers of Henry Clay: Supplement, 1793–1852* (Lexington, 1992), p. 281.
7. *The Horticulturist, and Journal of Rural Art and Rural Taste* 3 (1849) 233.
8. Corey Retter, *1861–1865 Union Executions* ( ), p. 169.
9. Gordon C. Roadarmel, 'Some California Dates of 1861', *California Historical Society Quarterly* 39 (1960) 295.
10. *Annals of Cleveland 1818–1935* (Cleveland, 1937), p. 49.
11. *Sullivan Democrat*, 16 May 1867.
12. *Augusta Chronicle* reprinted in the *Thomasville Daily Times Enterprise*, 1 October 1901; Riverside Louis P. Masur, *Rites of Execution: Capital Punishment and the Transformation of American Culture, 1776–1865* (Oxford, 1989), p. 97.
13. *Wellsville Allegany County Reporter*, 18 December 1889.
14. *Augusta Chronicle* reprinted in the *Thomasville Daily Times Enterprise*, 1 October 1901.
15. *Thirteenth Annual Report of the Officers of the Thirteen Club* (1895), p. 21.
16. Clifford R. Caldwell and Ron DeLord, *Eternity at the End of a Rope: Executions, Lynchings and Vigilante Justice in Texas 1819–1923* (Santa Fe, 2015), *passim*; *Fort Wayne Sentinel*, 19 April 1905; *Capital Punishment: Hearings Before the Committee on the Judiciary, United States Senate, Ninety-seventh Congress, First Session* (Washington, 1981), pp. 422, 573.
17. Guy Beiner, *Remembering the Year of the French: Irish Folk History and Social Memory* (Madison, 2007), pp. 217, 230.
18. L. Winstanley and H.J. Rose, 'Scraps of Welsh Folklore, I. Cardiganshire; Pembrokeshire', *Folklore* 37 (1926) 159. On urban development and the transience of ghost narratives see Karl Bell, 'Civic Spirits? Ghost Lore

and Civic Narratives in Nineteenth-Century Portsmouth', *Cultural and Social History* 18 (2014) 51–68.

19. *The Examiner*, 29 October 1845.
20. *Sheffield and Rotherham Independent*, 4 February 1868.
21. *Dundee Courier*, 27 March 1875.
22. *Wheeling Register*, 4 March 1888.
23. *Derry Journal*, 22 April 1908.
24. Charles Mackie, *Norfolk Annals: A Chronological Record of Remarkable Events in the Nineteenth Century* (Norwich, 1901), Vol. 1, entry 27th July 1807. Thanks to Elizabeth Hurren for this reference.
25. *St. James's Gazette*, 15 January 1883.
26. *London Evening Post*, 11 November 1749. On ghost hoaxers, see Davies, *The Haunted*.
27. *New York World*, 16 September 1894.
28. Brett Cogburn, *Rooster: The Life and Times of the Real Rooster Cogburn* (New York, 2012), pp. 92–93.
29. *Harrisburg Saline County Register*, 8 September 1899.
30. *Sedalia Weekly Sentinel*, 17 July 1903.
31. Beiner, *Remembering the Year*, p. 217.
32. Kathy Stuart, 'Suicide by Proxy: The Unintended Consequences of Public Executions in Eighteenth-Century Germany', *Central European History* 41 (2008), 413–445; Tyge Krogh, *A Lutheran Plague: Murdering to Die in the Eighteenth Century* (Leiden, 2012).
33. See, for example, Kirsi Kanerva, 'Having no Power to Return? Suicide and Posthumous Restlessness in Medieval Iceland', *Thanatos* 4, 1 (2015) 62–63; Nancy Caciola, 'Wraiths, Revenants, and Ritual in Medieval Culture', *Past and Present* 152 (1996) 3–45; Monballyu, *Six Centuries of Criminal Law*, pp. 231–234.
34. Riikka Miettinen and Evelyne Luef, 'Fear and Loathing? Suicide and the treatment of the corpse in early modern Austria and Sweden', *Frühneuzeit-Info* 23 (2012), 102; Stuart, *Defiled Trades*, p. 197.
35. Michael MacDonald and Terence R Murphy, *Sleepless Souls: Suicide in Early Modern England* (Oxford, 1990), pp. 44–49; Tarlow, *Ritual, Belief and the Dead*, pp. 147–149; Michael MacDonald, 'The Secularization of Suicide in England 1660–1800', *Past & Present* 111 (1986) 50–100.
36. *Portsmouth Evening News*, 14 December 1899; Davies, *The Haunted*, p. 52.
37. David Lederer, 'Living with the Dead: Ghosts in Early Modern Bavaria', in Kathryn A. Edwards (ed.), *Werewolves, Witches, and Wandering Spirits. Traditional Belief & Folklore in Early Modern Europe* (Kirksville, 2002), p. 37, fn. 46.
38. Stuart, *Defiled Trades*, p. 198.

39. Steve Roud, *The Penguin Guide to the Superstitions of Britain and Ireland* (London, 2003), pp. 91–92.

40. Milton Núñez, 'Remedies against Revenance: Two Cases from Old Hailuoto (Karlö), North Ostrobothnia, Finland', *Thanatos* 4, 2 (2015) 80; Alexander Kästner and Evelyne Luef, 'The Ill-Treated Body: Punishing and Utilizing the Early Modern Suicide Corpse', in Ward (ed.), *A Global History of Execution and the Criminal Corpse*, pp. 147–169. See also, Nicola Whyte, 'The deviant dead in the Norfolk landscape', *Landscapes* 4 (2003) 24–39.

41. Robert Halliday, 'The Roadside Burial of Suicides: An East Anglian Study', *Folklore* 121 (2010) 81–93; Halliday, 'Wayside graves and cross-road burials', *Norfolk Archaeology* 42 (1994) 80–83; Halliday, 'Wayside graves and crossroad burials' *Proceedings of the Cambridge Antiquarian Society* 84 (1995) 113–119.

42. 'Criminal Responsibility and Punishment for Suicide', *Central Law Journal* 55 (1902) 361. For examples of night burials, see: *The Huddersfield Chronicle and West Yorkshire Advertiser*, 19 November 1853; *Reynolds's Newspaper*, 19 May 1861; *The Bury and Norwich Post, and Suffolk Herald*, 23 December 1862; *Reynolds's Newspaper*, 14 January 1872; *Western Mail*, 16 February 1877; *Nottinghamshire Guardian*, 1 June 1889.

43. *Lancashire Daily Post*, 27 August 1906; *Lancashire Evening Post*, 16 January 1902.

44. Paul Hugger, 'Die Beerdigung der Selbstmörder im Kanton St. Gallen', *Schweizerisches Archiv für Volkskunde* 51 (1961), 41–48.

45. Milani, *Streghe*, p. 351; Gennaro Finnamore, *Tradizioni popolari abruzzesi* (Torino-Palermo, 1894), pp. 102–103.

46. G.J.C. Bois, *Jersey Folklore & Superstitions* (Milton Keynes, 2010), Vol. 1, pp. 16–17; *Notes & Queries* 99 (1851) 212. More generally, see Mark Norman, *Black Dog Folklore* (London, 2016).

47. See, for example, Paul Barber, *Vampires, Burial and Death: Folklore and Reality* (New Haven, 1988); David Keyworth, *Troublesome Corpses: Vampires and Revenants* (Southend-on-Sea, 2007); L'upcho S. Risteski, 'Categories of the "Evil Dead" in Macedonian Folk Religion', in Gábor Klanickzay and Éva Pócs (eds), *Christian Demonology and Popular Mythology* (Budapest, 2006), pp. 204–207.

48. Susan K. Morrissey, *Suicide and the Body Politic in Imperial Russia* (Cambridge, 2006), pp. 232–233.

49. David Lederer, 'The Dishonorable Dead: Elite and Popular Perceptions of Suicide in Early Modern Germany', in Sibylle Backmann, Hans-Jörg Künast, B. Ann Tlusty & Sabine Ullmann (eds), *Ehrekonzepte in der Frühen Neuzeit. Identität und Abgrenzungen* (Augsburg, 1998),

349–365; Evelyne Luef, 'Punishment Post Mortem—The Crime of Suicide in Early Modern Austria and Sweden', in Albrecht Classen and Connie Scarborough (eds), *Crime and Punishment in the Middle Ages and Early Modern Age* (Berlin, 2012), p. 570.

50. Paul Hugger, 'Die Beerdigung der Selbstmörder im Kanton St. Gallen', *Schweizerisches Archiv für Volkskunde* 51 (1961) 41–48. For general ideas about the German tradition of suicide burials, see also: Paul Geiger, 'Die Behandlung Der Selbstmörder im deutschen Brauch', *Schweizer Volkskunde* 26 (1925–1926) 145–160.

51. Thomas A. Kselman, *Death and the Afterlife in Modern France* (Princeton, 1993), pp. 104–105.

52. Morrissey, *Suicide and the Body Politic*, p. 233.

53. Giuseppe Nicasi, 'Le credenze religiose delle popolazioni rurali dell'alta valle del Tevere', *Lares* 1 (1912) 167–169; Pietrina Moretti, 'Ora feriada e ora mala', *Lares* 21 (1955) 61–64; Alfonso Maria Di Nola, *La Nera Signora: antropologia della morte* (Rome, 1995), p. 128; Giovanni Tassoni, *Arti e tradizioni popolari. Le inchieste napoleoniche sui costumi e le tradizioni nel regno italic* (Bellinzona, 1973), pp. 134, 147.

54. Marisa Milani, *Streghe, morti ed esseri fantastici nel Veneto* (Padova, 1994), pp. 350–351.

55. Giovanni Bronzini, *Tradizioni popolari in Lucania* (Matera, 1953), pp. 243–244.

56. Maria Pia Di Bella, 'Pietà e Giustizia. La "santificazione" dei criminali giustiziati', *La Ricerca Folklorica* 29 (1994) 69–72; Maria Pia Di Bella, 'Conversion and Marginality in Southern Italy', in Andrew Buckster and Stephen D. Glazier (eds), *The Anthropology of Religious Conversion* (Lanham, 2003), pp. 85–87.

57. Michael P. Carroll, *Veiled Threats: The Logic of Popular Catholicism in Italy* (Baltimore, 1996), pp. 148–151. See also, Adriano Prosperi, 'Il sangue e l'anima: ricerche sulle compagnie di giustizia in Italia', *Quaderni storici* 17 (1982) 959–999.

58. E. Sidney Hartland, 'The Cult of Executed Criminals at Palermo', *Folklore* 21 (1910) 168–179; Maria Tedeschi, 'Canti di devozione alle Anime del Purgatorio e dei Corpi decollati', *Archivio per lo studio delle tradizioni popolari* 16 (1937); Giuseppe Pitrè, *Usi e costumi credenze e pregiudizi del popolo siciliano* (Palermo, 1978) Vol. 4, pp. 8–11.

59. Giuseppe Pitrè, *Biblioteca delle tradizioni popolari* 24 (1913), 185–186; Binde, *Bodies of Vital Matter*, p. 125.

60. Hartland, 'The Cult', 174–175.

61. Giuseppe Pitrè, *Biblioteca*, Vol. 24, 189–190; Sebastiano Salomone, *Storia di Augusta* (Siracusa, 1905), p. 295.

62. Giuseppe Pitrè, *Biblioteca*, Vol. 24, p. 187.

63. Cited in Carroll, *Veiled Threats*, p. 156.
64. Binde, *Bodies of Vital Matter*, p. 127.
65. Tausiet, *Urban Magic*, p. 149.
66. Emma Hardinge, *Modern American Spiritualism* (New York, 1870), p. 285.
67. J.G.H. Brown, *A Message from the World of Spirits* (London, 1857).
68. See Logie Barrow, *Independent Spirits: Spiritualism and English Plebeians, 1850–1910* (London, 1986), Ch. 3.
69. Brown, *Message*, pp. 172–173.
70. See Georgina Byrne, *Modern Spiritualism and the Church of England, 1850–1939* (Woodbridge, 2010), pp. 90–91.
71. Francis H. Smith, *My Experience: Or, Foot-Prints of a Presbyterian to Spiritualism* (Baltimore, 1860), pp. 132–133.
72. Christine Ferguson, *Determined Spirits: Eugenics, Heredity and Racial Regeneration in Anglo-American Spiritualist Writing, 1848–1930* (Edinburgh, 2012), pp. 154–156; James Gregory, *Victorians against the Gallows: Capital Punishment and the Abolitionist Movement in Nineteenth Century Britain* (London, 2012), Ch. 4.
73. Newton Crosland, *Apparitions; A New Theory* (London, 1856), p. 20. See also, 'The Literature of Spirit-Rapping', *The National Review* 4 (1857) 142.
74. William Q. Judge, *The Ocean of Theosophy* (New York, [1893] 1910), pp. 108, 48.
75. Eliab Wilkinson Capron, *Modern Spiritualism: Its Facts and Fanaticisms* (Boston, 1855), pp. 114–117. For details of the Adler murder, see: https://privatelettersjsg.wordpress.com/2014/06/27/1849-murder-of-nathan-adler/; *Particulars of the murder of Nathan Adler on night of November sixth, 1849* (New York, 1850).
76. William R. Gordon, *A Three-Fold Test of Modern Spiritualism* (New York, 1856), pp. 40–42.
77. *The Medium and Daybreak*, 24 July 1885, 466. See also, the Gerald Massey archive site http://gerald-massey.org.uk/massey/cbiog_part_06.htm.
78. Brown, *Message from the World of Spirits*, p. 34.

INDEX

© The Editor(s) (if applicable) and The Author(s) 2017
O. Davies and F. Matteoni, *Executing Magic in the Modern Era*,
Palgrave Historical Studies in the Criminal Corpse and its Afterlife,
DOI 10.1007/978-3-319-59519-1